Synthesis Lectures on Chemical Engineering and Biochemical Engineering

This series publishes short books on all aspects of chemical engineering, covering the analysis or design of chemical processes to effectively convert materials into more useful materials or energy. The books will focus on fundamental aspects necessary for chemical engineering design including chemistry, math, physics, and sometimes biology to improve the quality of life by inventing, optimizing, and economizing new technologies and products.

Navnath Hatvate · Hemantkumar N. Akolkar ·
A. K. Haghi

Sonochemistry

A Sustainable Technology

Navnath Hatvate
Institute of Chemical Technology
Mumbai, Marathwada Campus, Jalna
Maharashtra, India

A. K. Haghi
Department of Chemistry
Institute of Molecular Sciences
University of Coimbra
Coimbra, Portugal

Hemantkumar N. Akolkar
Department of Chemistry
Rayat Shikshan Sanstha's Abasaheb Marathe
Arts and New Commerce Science College
Rajapur, Maharashtra, India

ISSN 2327-6738 ISSN 2327-6746 (electronic)
Synthesis Lectures on Chemical Engineering and Biochemical Engineering
ISBN 978-3-031-91655-7 ISBN 978-3-031-91656-4 (eBook)
https://doi.org/10.1007/978-3-031-91656-4

© The Editor(s) (if applicable) and The Author(s), under exclusive license to Springer
Nature Switzerland AG 2026

This work is subject to copyright. All rights are solely and exclusively licensed by the Publisher, whether the whole or part of the material is concerned, specifically the rights of translation, reprinting, reuse of illustrations, recitation, broadcasting, reproduction on microfilms or in any other physical way, and transmission or information storage and retrieval, electronic adaptation, computer software, or by similar or dissimilar methodology now known or hereafter developed.
The use of general descriptive names, registered names, trademarks, service marks, etc. in this publication does not imply, even in the absence of a specific statement, that such names are exempt from the relevant protective laws and regulations and therefore free for general use.
The publisher, the authors and the editors are safe to assume that the advice and information in this book are believed to be true and accurate at the date of publication. Neither the publisher nor the authors or the editors give a warranty, expressed or implied, with respect to the material contained herein or for any errors or omissions that may have been made. The publisher remains neutral with regard to jurisdictional claims in published maps and institutional affiliations.

This Springer imprint is published by the registered company Springer Nature Switzerland AG
The registered company address is: Gewerbestrasse 11, 6330 Cham, Switzerland

If disposing of this product, please recycle the paper.

Preface

Sonochemistry, which examines the impact of ultrasound on chemical systems, has become an effective instrument for advancing sustainable technology. In recent decades, sonochemistry has transitioned from a specialized research domain to a dynamic field with diverse applications. This book provides an overview of sonochemistry's principles, applications, and future directions.

The book looks at the basic ideas of sonochemistry, mainly the effects of ultrasound on matter and energy, as well as acoustic cavitation and sonochemical reactions. The primary motivation for this book is to emphasize the potential of sonochemistry as a sustainable technology. Sonochemistry presents several advantages compared to conventional chemical processes, such as decreased energy consumption, enhanced reaction rates, and superior product yields. Sonochemistry can synthesize various materials, such as nanoparticles, nanocrystals, and other advanced materials.

The book comprises five chapters, each addressing a distinct aspect of sonochemistry. Chapter 1 introduces the essential principles of sonochemistry, detailing ultrasound's physical and chemical effects. Chapter 2 examines the applications of sonochemistry within the field of chemistry, focusing on synthesizing organic compounds and preparing nanomaterials using the sonochemical approaches. Chapter 3 examines the applications of sonochemistry within chemical engineering, specifically addressing the design of sonochemical reactors and the optimization of sonochemical processes. Chapter 4 examines the applications of sonochemistry within environmental engineering, focusing on the remediation of contaminated soil and groundwater. Chapter 5 presents an overview of future sonochemistry directions, focusing on combining sonochemical technologies and investigating new applications in pharmaceutical and other industries.

This book targets researchers, students, and professionals engaged in sonochemistry and its applications. This book aims to be a valuable resource for understanding the principles and applications of sonochemistry and contributing to the advancement of sustainable technologies.

In summary, sonochemistry represents a dynamic area of study with diverse applications. This book thoroughly examines the principles, applications, and future directions of sonochemistry, serving as a valuable resource for researchers, students, and professionals in the field.

Jalna, India	Dr. Navnath Hatvate
Rajapur, India	Dr. Hemantkumar N. Akolkar
Edinburgh, UK	Dr. A. K. Haghi

Contents

1	**Sonochemistry Science and Engineering: Understanding the Concepts**	**1**
1.1	Introduction	1
1.2	Acoustic Cavitation	4
1.3	The Chemistry of Ultrasound	7
1.4	Ultrasound-Assisted Approaches	9
	1.4.1 Homogeneous Sonochemistry	10
	1.4.2 Heterogeneous Sonochemistry	11
	1.4.3 Dual-Frequency Ultrasound	12
	1.4.4 Ultrasound in Microreactors	13
1.5	Sonochemical Catalysis	14
1.6	Conclusion	15
	References	15
2	**Applications in Chemistry**	**21**
2.1	Introduction	21
2.2	Ultrasound in Synthetic Organic Chemistry	22
	2.2.1 Homogeneous Sonochemistry	23
	2.2.2 Heterogeneous Sonochemistry	27
	2.2.3 Sonocatalytic Reaction	33
2.3	Organic Sonochemistry: Mechanism and Reactivity	36
	2.3.1 Cavitation and Radical Formation	36
	2.3.2 Mechanical Effects	37
	2.3.3 Reactivity and Sonochemical Switching	37
2.4	Application in Preparation of Inorganic Nanomaterials	38
	2.4.1 Metal Oxide Nanoparticles	39
	2.4.2 Silicas and Organosilica	39
	2.4.3 Metal–Organic Frameworks (MOFs) and Covalent-Organic Frameworks (COFs)	40

		2.4.4 Carbons	41
	2.5	Conclusion	42
	References		43

3 Sonochemistry in Chemical Engineering ... 47

- 3.1 Introduction ... 47
- 3.2 Sonochemical Reactors ... 49
  - 3.2.1 Ultrasonic Horn Reactors ... 49
  - 3.2.2 Ultrasonic Bath Reactors ... 51
  - 3.2.3 Multiple-Frequency Flow Cell Reactors ... 52
- 3.3 Bubble Dynamics ... 52
  - 3.3.1 Mechanism of Bubble Cavitation ... 54
- 3.4 Factors Affecting Cavitation ... 56
  - 3.4.1 Ultrasonic Frequency ... 56
  - 3.4.2 Ultrasonic Intensity ... 57
  - 3.4.3 Liquid Properties ... 58
  - 3.4.4 Temperature ... 59
  - 3.4.5 Pressure ... 60
  - 3.4.6 Reactor Design and Transducer Placement ... 61
- 3.5 Effects of Ultrasound on Chemical Systems ... 62
  - 3.5.1 Solid–Liquid Systems ... 62
  - 3.5.2 Liquid–Liquid Systems ... 64
  - 3.5.3 Gas–Liquid Systems ... 65
  - 3.5.4 Reaction Processes in Chemical Engineering ... 66
- 3.6 Modeling of Mass Transfer Effect ... 67
  - 3.6.1 Sonochemical Factors Affecting Mass Transfer ... 67
  - 3.6.2 Transfer Correlations for Ultrasonic Reactors ... 68
  - 3.6.3 Diffusion-Limited Model for Vapor Transport in Cavitation Bubbles ... 69
- 3.7 Methods Used to Produce Ultrasound ... 71
  - 3.7.1 Piezoelectric Transducers ... 71
  - 3.7.2 Magnetostrictive Transducers ... 72
  - 3.7.3 Ultrasonic Horns or Probes ... 72
  - 3.7.4 Laser-Generated Ultrasound ... 72
  - 3.7.5 Electrostatic or Electroacoustic Methods ... 73
- 3.8 The Control of Airborne Contamination ... 73
- 3.9 Wastewater Treatments ... 74
- 3.10 Energy Consumption Control for Chemical Transformations ... 77
- 3.11 Conclusion ... 78
- References ... 78

4	**Applications in Environmental Engineering**		91
	4.1	Introduction	91
	4.2	Environmental Sonochemistry	93
	4.3	Elimination of Hazardous Substances	94
	4.4	The Use of Less Hazardous Chemicals and Environmentally Friendly Solvents	98
	4.5	Environmental Remediation	100
		4.5.1 Sonochemistry in Wastewater Treatment	100
		4.5.2 Sonochemistry in Sludge Treatment	102
	4.6	Pollution Prevention	102
	4.7	Purification of Water	104
	4.8	Decontamination of Soil	105
	4.9	Conclusion	106
	References		107
5	**Future Outlook on Sonochemistry**		113
	5.1	Introduction	113
	5.2	Sonochemistry and Pharmaceutical Sciences	114
	5.3	Sonochemistry and Microwaves	116
	5.4	Industrial Sonochemistry	117
	5.5	Conclusion	118
	References		119

About the Authors

Navnath Hatvate is an assistant professor at the Institute of Chemical Technology, Mumbai, Marathwada Campus, Jalna. He holds a Master of Pharmacy and a Ph.D. from the Department of Pharmaceutical Sciences and Technology at the Institute of Chemical Technology, Mumbai. His areas of expertise and research interests include the development of new chemical entities (NCE) and their analogues, modifications of excipients and their applications in drug delivery, the use of AI and ML in drug discovery and drug delivery, process chemistry of APIs and intermediates, and total synthesis of natural products, bio-active compounds, application of novel technologies in organic synthesis. Dr. Hatvate has published over 50 research publications and authored 4 books. Currently he serve as a editorial board member for Scientific Reports (Nature Publishing).

Hemantkumar N. Akolkar, Ph.D. is currently working as an Associate Professor in the Department of Chemistry, Abasaheb Marathe College, Rajapur, Maharashtra, India. He is an acclaimed academician and researcher who has worked for the last 14 years at both undergraduate and postgraduate levels. He has published 55+ research papers in national and international journals and conferences of repute. His areas of interest are heterocyclic chemistry, synthetic organic chemistry, and green chemistry. He is a reviewer of several journals of international repute in chemistry.

A. K. Haghi is Professor Emeritus of Engineering Sciences and has published 250+ academic research-oriented books and 1000+ research papers in various journals and conference proceedings. His leadership in academic publishing includes founding and serving as Editor-in-Chief of prestigious journals. Professor Haghi's extensive educational background and supervisory roles underscore his expertise and contributions to the field of engineering sciences. He is currently appointed as Honorary Research Associate (HRA) at University of Coimbra, Portugal.

List of Figures

Fig. 1.1	Timeline for development of sonochemistry	3
Fig. 1.2	Stages of acoustic cavitation: bubble formation, growth, and collapse	5
Fig. 2.1	Preparation of nanomaterials using sonochemistry (created by biorender.com)	42
Fig. 3.1	Vertical ultrasonic horn reactors and horizontal ultrasonic horn reactors (created using Biorender.com)	50
Fig. 3.2	Ultrasonic bath reactors (created using Biorender.com)	51
Fig. 3.3	Multiple-frequency flow cell reactors (created using Biorender.com)	53
Fig. 4.1	US combined with wastewater treatment plant (created using Biorender.com)	105
Fig. 5.1	Ever-increasing boon of research in sonochemistry (created using Biorender.com)	114
Fig. 5.2	US combined with MW reactor (created using Biorender.com)	117

List of Tables

Table 3.1	Prevention of airborne contamination given by Aslam et al. (2024)	75
Table 4.1	Different sonochemical methods used for the degradation of environmental pollutants	95
Table 5.1	List of nanoformulations	116
Table 5.2	List of industries using the sonochemical approach	118

Sonochemistry Science and Engineering: Understanding the Concepts

Abstract

Sonochemistry, which involves acoustic cavitation with the help of ultrasonic waves, is an emerging field in every domain. As known to all, the field finds wide applications in pharmaceuticals, polymer sciences, dentistry, mining, extraction, and many others. This field can also prove to be a boon for the environment due to its vast applications in environmental remediation, removal of pollutants, using environmentally friendly methodology, etc. Sonochemistry is a recent area of interest due to its low cost, reduced waste generation, shorter time requirement, and increased efficiency of processes. This chapter gives insights into the development of ultrasound, its application, concept, and challenges.

1.1 Introduction

The use of ultrasound can be traced back to the early twentieth century, but it was first applied in the medical field and sonar (Newman and Rozycki 1998). One of those inventions was carried out during the First World War by French Physicist Paul Langevin; he made the initial step for ultrasonic applications that would benefit imaging and material assessment in the subsequent decades (Duck and Thomas 2022). The 1980s were a pivotal year as ultrasound entered the facet of chemical engineering, primarily due to efforts such as those of Suslick, who worked on the nature of cavitation (Rosales Pérez and Esquivel Escalante 2024). This process in which microbubbles are formed and then burst under ultrasonic waves to generate high-temperature environments was appreciated for their capability of achieving chemical transformations that are difficult under normal circumstances due to the extreme conditions, such as temperatures of over 5000 K and pressures exceeding 1000 atm (Mason 2009). These studies led to the rapid development of sonochemistry, and researchers such as TJ Mason and D. Peters developed the theory and applied aspects of the phenomenon (Mohamed et al. 2016).

Ultrasound has a frequency ranging from 20 kHz to a few MHz and industrial applications use high or low intensity depending on the process (Dehghani et al. 2010). The movement of ultrasound through a medium sets up compressional and tensile waves, which form the high-energy regions essential for processes such as cavitation (Humphrey 2007). The most crucial of sonochemistry is cavitation, as the collapsing microbubbles create extremely localized conditions that lead to chemical reactions (Suslick 2003). Lord Rayleigh scientifically and mathematically explained this in the nineteenth century, forming the basis of all ultrasonic applications in chemical engineering (Pokhrel et al. 2016).

Compared with conventional practice in chemical engineering, ultrasound has many benefits as a disruptive technology. It optimizes process efficacy, reduces energy utilization, and optimizes wastage (Bhargava et al. 2021). It must be pointed out that the process is based on principles other than thermal or mechanical energy, making it more sustainable than traditional methods: ultrasound can cause reactions under non-equilibrium conditions (Safwa et al. 2023). For example, ultrasonic reactors enhance the mixing and dispersion attributes to improve reaction yields and selectivity (Adamou et al. 2024). Furthermore, ultrasound is sensitive to cold climates or low pressure, so its power consumption is comparatively low (Yao et al. 2020). It encompasses a wide range of areas, such as crystallization, polymerization, and water treatment, which raises its applicability to solving numerous engineering issues (Singla and Sit 2021).

Nowadays, environmental resource requirements have increased, affecting nature overall; thus, utilizing the available resources and finding other safe alternatives for environment remodeling is essential. Sonochemistry is one such area that satisfies the needs of humans without causing harm to the environment. Alternatively, it is currently one of the greener technologies available (Meng et al. 2015). Ultrasound interacts with liquids at frequencies to produce the peculiar phenomenon known as acoustic cavitation. Cavitation is the creation, expansion, and explosion of gaseous bubbles in the liquid phase. Sonochemistry has wide applications in improving reaction speed, altering reaction routes, and enabling new chemical transformations. Sonochemistry has applications in many industries, including food processing, medicines, environmental remediation, and material synthesis. Green chemistry principles are aligned with the use of ultrasounds in chemical processes since they speed up reactions and make cleaner, greener alternatives possible by lowering the demand for harsh reagents and conditions (Mason and Lorimer 2002).

The engineering side of sonochemistry aims to optimize cavitation efficiency through reactor design and process parameter optimization. Various reactor types, including batch, flow, and hybrid systems, have been developed to meet specific industrial demands. Research on the dynamics of cavitation and how it interacts with various chemical systems

1.1 Introduction

is still crucial for bridging the gap between lab-scale studies and industrial-scale applications. Additionally, sonochemistry influences the development of novel materials and processes in cutting-edge fields, including biotechnology, nanotechnology, and renewable energy (Ashokkumar 2011).

The basic principle of sonochemistry follows free radical formation due to cavitation, which occurs due to the rarefaction of sound waves. In 1994, Harvey discovered the term rectified diffusion, which gave rise to the concept of bubble oscillation due to mass transfer. During the 1950s, significant evolution happened in the field of sonochemistry, which caused enormous development in this field. Noltingk and Neppiras, in 1950, gave the first computerized calculations in the bubble. Afterward, in 1953, Schultz and Henglein gave the sonolysis of liquids. In 1954, Elder suggested that one of the elements causing the well-known ultrasonic cleansing effects in diverse environments was bubble-induced microstreaming (Thompson and Doraiswamy 1999).

Figure 1.1 represents a timeline illustrating sonochemistry's historical evolution and milestones. It visually depicts the progression of discoveries, technological advancements, and the recognition of sonochemistry's significance as a scientific discipline over time. Key breakthroughs and pivotal events are marked along the timeline to highlight their impact on the development of ultrasound applications and sonochemistry research (Carlton 2019).

Cavitation leads to an explosion of bubbles, which allows for the formation of low-pressure areas. As the sound waves keep reaching, these bubbles increase in size until they cannot grow anymore, leading to sudden collapse. This explosion is enough to break chemical bonds or clean surfaces at a microscopic level. Sonochemistry extensively uses this phenomenon to clean, decompose supplies, and initiate chemical processes (Adewuyi 2005). Acoustic cavitation produces high-pressure and temperature areas, which starts several chemical and biological reactions. Based on its principle, it has better advancements

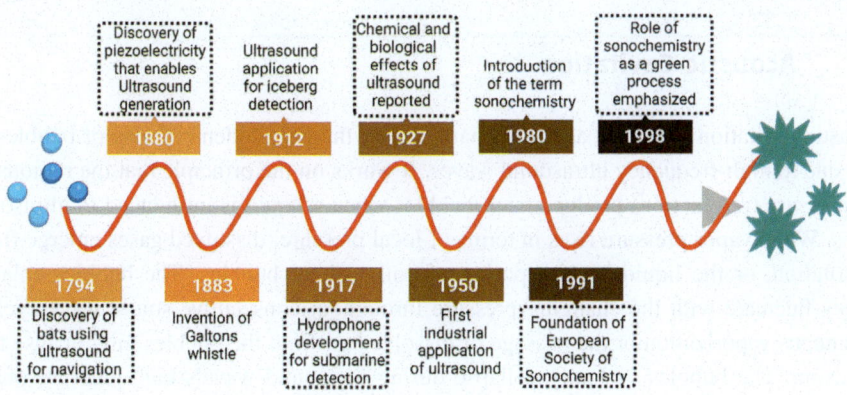

Fig. 1.1 Timeline for development of sonochemistry

in food processing (Hoo et al. 2022), material engineering (Neppolian et al. 2012), and electrochemistry (Gadge et al. 2024). The number of continuous cavities that form and collapse at different times makes exploitation possible. Until recently, the only method for creating cavities that have been thoroughly researched was ultrasound (Kasprzyk-Hordern et al. 2003).

Sonochemistry, induced by ultrasound, can be understood through its interaction with different phase systems. In liquid–gas systems, the mechanism of acoustic cavitation is followed, as mentioned earlier. This phenomenon creates localized zones of high pressure and temperature, leading to intense interactions between the gas and liquid phases, as observed in processes driven by high-speed flows or ultrasonic waves (Zigangareeva and Kiselev 1998). In liquid–liquid systems, ultrasound facilitates the mixing and interaction of immiscible liquids by producing fine dispersions and stable emulsions. This enhances chemical reactions, improves mass transfer, and significantly reduces reaction times, making ultrasound highly beneficial (Li et al. 2021). In liquid–solid systems, the effects of ultrasound can be categorized into those involving cavitation and those without. Cavitation effectively cleans solid surfaces by removing and degreasing impurities while promoting reactions like metal dissolution, oxidation, and catalytic processes. It effectively works in crucial chemical and petrochemical processes, including hydrogenation and hydrocarbon reforming. Without cavitation, ultrasound still enhances mass transfer between solid and liquid phases by facilitating the movement of particles or molecules, increasing reaction efficiency in the absence of bubble collapse. The use of power ultrasound, particularly in the 2–10 kHz range, has become effective in sonochemistry, offering a unique energy source distinct from traditional methods such as heat, light, or pressure. This energy enables the formation of microbubbles during the rarefaction phase of ultrasound waves, which collapse under compression to produce localized extreme conditions, such as temperatures reaching thousands of kelvin and pressures of several atmospheres, thereby driving chemical transformations with remarkable efficiency (Margulis 1985).

1.2 Acoustic Cavitation

Acoustic cavitation is termed as the formation, growth, and sudden collapse of bubbles in fluid due to high-frequency ultrasound waves. It works on the principle that the regions of low pressure (rarefaction) produce microbubbles when a liquid is introduced to ultrasonic waves. When vapor pressure rises in terms of local pressure, dissolved gases emerge from the solution, or the liquid itself vaporizes, forming these bubbles. The bubbles enlarge as they fluctuate with the changing pressure throughout consecutive sound wave cycles. Continuous vaporization or the passage of dissolved gas into the bubbles might cause this expansion. The bubbles violently collapse during the sound wave's high-pressure phase (compression) once they reach a crucial size, forming an intensified collapse condition (Neppiras 1980). Acoustic cavitation is a fundamental process in sonochemistry, where

1.2 Acoustic Cavitation

chemical reactions are initiated or accelerated by the tremendous energy released during bubble collapse. In aqueous systems, the explosion of bubbles can produce reactive radicals, such as hydroxyl radicals, which can fuel oxidation processes and other chemical changes. Additionally, this technique produces "hot spots" where localized energy concentrations might improve reaction speeds without requiring heating. However, because these harsh circumstances are temporary and limited to microscopic areas, sonochemistry is a very effective and selective method (Suslick 1990). Several variables affect cavitation dynamics, including the ultrasound's frequency and strength, the liquid's characteristics, and the existence of dissolved gasses or other contaminants. For instance, larger bubbles that explode more forcefully at lower frequencies tend to release more energy. Conversely, smaller, more stable bubbles that collapse with less energy are produced at higher frequencies, which may offer greater control over response pathways. Furthermore, the properties of the liquid affect the cavitation threshold or the lowest ultrasonic intensity required to cause bubble formation (Guo and Zhu 2018).

Figure 1.2 illustrates the fundamental stages of acoustic cavitation driven by ultrasonic waves. It begins with the formation of cavitation bubbles, a result of the rapid compression and rarefaction cycles induced by the propagation of sound waves in a liquid medium. Further, the growth and enlargement of the bubbles occur as the surrounding fluid pressure decreases, allowing gas and vapor to enter the bubble during successive rarefaction cycles. Finally, the collapse of the bubbles, where the sudden implosion generates intense localized conditions, including extremely high temperatures and pressures. This collapse produces highly reactive species such as hydroxyl and hydrogen radicals, making the process pivotal for sonochemical reactions (Carlton 2019).

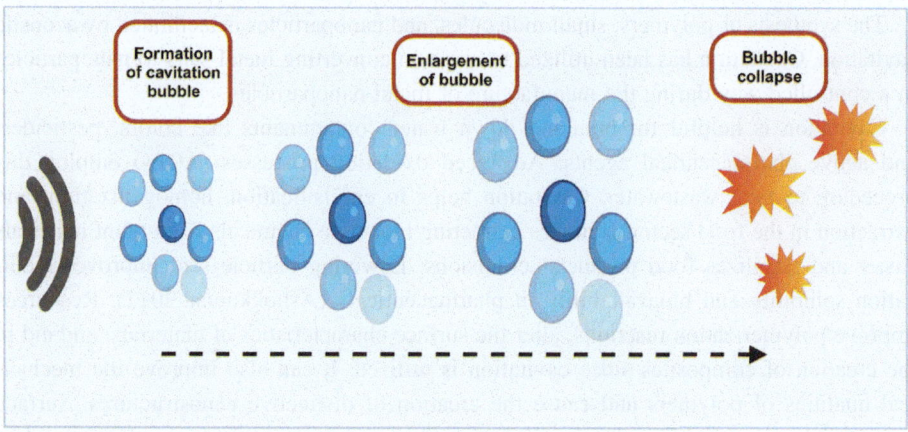

Fig. 1.2 Stages of acoustic cavitation: bubble formation, growth, and collapse

The three main phases of the acoustic cavitation mechanism are bubble nucleation, growth, and collapse. High- and low-pressure regions are produced by alternating rarefaction cycles and compression when an ultrasonic wave travels through a liquid. Microbubbles or cavities form when the pressure falls below the liquid's vapor pressure during rarefaction. This process, called nucleation, can occur around contaminants that function as nuclei or molecules of dissolved gas. These bubbles enlarge throughout growth, taking in energy from later rarefaction cycles. The frequency and strength of the ultrasound will determine how many sonic cycles this growth takes place. Additionally, the collapse develops high-speed liquid jets and shock waves that improve mass transfer and can physically shatter surfaces or particles. In aqueous settings, this powerful energy release causes water molecules to produce reactive species like hydroxyl radicals ($OH^.$) and hydrogen radicals ($H^.$), which start chemical processes. These radicals can drive oxidation and reduction reactions, making their creation essential in sonochemical processes (Gogate and Pandit 2004a, b).

Cavitation is either stable (non-inertial) or transient (inertial). Sonoluminescence, or the emission of light from collapsing bubbles, is frequently the result of transient cavitation, which is characterized by bubbles that expand quickly and collapse violently. In contrast, stable cavitation produces microstreaming effects (Mason and Lorimer 2002). Controlling cavitation is crucial for optimizing sonochemical processes. Variables like reactor design, ultrasonic power, and solvent selection are essential to modify cavitation behavior. Scientists can adjust reactions to achieve specific results by comprehending these variables, including increasing yield, selectivity, or reaction time. Advanced sonochemical reactors, such as dual-frequency and focused ultrasound systems, have been developed to expand the possibilities of cavitation-driven processes in various industries, including environmental remediation and material synthesis (Leighton and Apfel 1994).

The synthesis of polymers, small molecules, and nanoparticles is facilitated by acoustic cavitation. Cavitation has been utilized to assist in converting metal salts to nanoparticles in a controlled way during the manufacture of metal nanoparticles.

Cavitation is helpful for breaking down water contaminants like colors, pesticides, and active pharmaceutical agents. Advanced oxidation processes (AOPs) employ this procedure to treat wastewater. Cavitation helps in emulsification, homogenization, and extraction in the food sector. It makes extracting bioactive chemicals from plant materials easier and stabilizes food products' emulsions. Lowering particle size improves medication solubility and bioavailability in pharmaceuticals (Ashokkumar 2011). Resources improve polymerization reactions, alter the surface characteristics of materials, and aid in the creation of composites since cavitation is utilized. It can also improve the mechanical qualities of polymers and cause the creation of distinctive nanostructures. Surface preparation and cleaning cavitation is frequently employed in ultrasonic cleaning when impurities are removed from surfaces by the powerful shock waves and jets produced

1.3 The Chemistry of Ultrasound

during bubble collapse. Applications like these are essential in fields like electronics and medical device manufacture that demand high purity (Esclapez et al. 2011).

Ultrasound represents a transformative tool in modern chemistry, leveraging high-frequency sound waves to induce significant physical and chemical changes in various media. The fundamental mechanism underlying these effects is acoustic cavitation, a phenomenon driven by the wave nature of sound. Ultrasound propagates through liquids as alternating compression and rarefaction waves. During the rarefaction phase, regions of low pressure are created, causing the liquid to stretch and form microbubbles when the pressure falls below its vapor pressure. These bubbles grow over successive cycles and eventually collapse violently under the compression phase, releasing vast amounts of energy locally (Athanassiadis et al. 2022). As sound waves travel through a liquid medium, their oscillatory nature generates localized high-energy zones. This oscillation facilitates the periodic formation and collapse of bubbles, with each cycle amplifying the intensity of the implosion. The collapse of these microbubbles generates extreme localized conditions, with temperatures exceeding 5000 K and pressures reaching several hundred atmospheres. These transient but intense conditions provide a fertile ground for chemical transformations, often unattainable under conventional conditions. The chemistry of ultrasound also extends to its ability to enhance mass transfer and mixing in heterogeneous systems. Cavitation-induced microstreaming increases the interaction between reactants and catalysts, improving chemical transformations' efficiency and selectivity (Leong et al. 2016).

One of the most critical chemical outcomes of cavitation is the generation of hydroxyl radicals and hydrogen radicals through the homolytic cleavage of water molecules. When cavitation bubbles collapse, the energy released is sufficient to break the bonds within water molecules (Reaction Scheme 1.1):

The extreme conditions within the bubble, with temperatures exceeding 5000 K and pressures reaching several hundred atmospheres, drive this bond dissociation process. The hydroxyl radicals generated are powerful oxidizing agents, while hydrogen radicals act as reducing agents. These radicals serve as the primary reactive intermediates in many sonochemical reactions, contributing to processes such as oxidation, reduction, and polymerization (Zupanc et al. 2019).

Reaction Scheme 1.1
Hydrolysis of water molecule under ultrasonic irradiations

$$H_2O \xrightarrow{\text{Cavitation}} OH\cdot + H\cdot$$

The radical formation occurs in "hot spots," localized zones within the collapsing bubbles where energy is concentrated. These hot spots exist for only microseconds, but their extreme conditions allow rapid and selective chemical transformations. Unlike traditional heating methods, where the bulk temperature of the liquid rises, the energy in ultrasonic cavitation is confined to these microscopic zones, leaving the bulk of the liquid relatively unaffected. This feature makes ultrasound chemistry more energy-efficient and environmentally friendly, as it minimizes the risk of degrading sensitive molecules in the surrounding liquid. For example, hydroxyl radicals react with organic molecules to produce oxidized products in oxidation reactions. In contrast, hydrogen radicals facilitate the reduction of functional groups such as nitro to amine in reduction reactions. These radicals' controlled generation and localization make cavitation a versatile tool for chemical synthesis (Wu et al. 2020).

The radicals formed during cavitation can participate in a variety of secondary reactions, either recombining or reacting with other species in the liquid. For instance,

Reaction Scheme 1.2 forms hydrogen peroxide, a strong oxidant widely used in wastewater treatment and organic synthesis of molecular hydrogen, which has applications in hydrogenation processes and renewable energy systems (Yusof et al. 2016).

The formation of radicals during cavitation can be understood through quantum mechanical principles, where the energy released during bubble collapse exceeds the bond dissociation energy of water molecules. This energy is sufficient to break the O–H bonds in water, producing radicals with unpaired electrons. These highly unstable radicals seek to react with other molecules or radicals to attain stability, driving a cascade of chemical reactions. The radicals generated during cavitation can synergize with catalysts to enhance reaction rates and selectivity. For example, using solid catalysts in conjunction with ultrasound improves the efficiency of hydrogenation and oxidation reactions by increasing the interaction between the radicals and the catalyst surface. Ultrasound is wave-driven energy transfer, and its ability to create non-equilibrium conditions underscores its unique role in chemistry. By combining physical wave dynamics with chemical reactivity, ultrasound enables sustainable and efficient processes, aligning with green chemistry principles. Its applications in diverse fields, from environmental remediation to material synthesis, highlight this versatile tool's transformative potential (Yasui 2022).

Piezoelectric transducers are majorly used to produce ultrasound, in which the electrical charges are produced by some forms of solid materials and converted into energy (Pokhrel et al. 2016). The frequency produced by the transducer plays an essential role

Reaction Scheme 1.2 Generation of reactive intermediates from hydroxy and hydrogen radicals

$$OH\cdot + OH\cdot \longrightarrow H_2O_2$$

$$H\cdot + H\cdot \longrightarrow H_2$$

in applying sonochemistry (Draye and Kardos 2016). The synthesis and alteration of organic substances have been regarded as among the most significant contemporary uses of sonochemistry. High-intensity ultrasonography has a variety of uses in many systems, primarily because of the cavitation effect. Through the creation of severe conditions in the medium, ultrasound improves a variety of chemical and physical reactions that can break into pollutants, synthesize nanomaterials, emulsions, cleaning, and catalysis in a variety of industries, including electronics, food, pharmaceuticals, and water treatment (Li et al. 2021).

Ultrasound in chemistry primarily leverages the effects of cavitation and sound wave-induced motion to accelerate process efficiency. Compared to conventional chemistry, ultrasound interacts with sonic cavitation rather than directly with chemical bonds. (Gogate and Pandit 2004a, b). The bubble dynamics mechanism is essential to sonochemistry (see Chap. 3). Besides the above applications, ultrasound facilitates extraction, emulsification, and homogenization procedures in the food and pharmaceutical sectors. It promotes emulsion stability and the extraction of bioactive components from plant sources. Ultrasound is optimized for particular purposes using various methods and reactor designs. Microreactors and flow reactors are also becoming more popular in sonochemistry. Precision synthesis is made possible by these reactors because they provide continuous processing and improved control over reaction conditions. Because of their high surface-to-volume ratios and small volumes, microreactors improve cavitation effects, which boosts the productivity of chemical synthesis and pharmaceutical manufacturing (Mehta et al. 2022).

1.4 Ultrasound-Assisted Approaches

The capacity of ultrasound-assisted methods to develop acoustic cavitation, which causes physical and chemical changes, has transformed many facets of chemistry and engineering. Numerous chemical reactions and operations that would typically need severe conditions or be impossible to accomplish using traditional methods are made possible by these extreme microenvironments (Mason and Lorimer 2002). When bubbles collapse in sonochemistry, localized hot spots are created that operate as microreactors, promoting bond breaking and the generation of free radicals. For instance, ultrasonication can dramatically increase reaction speeds and selectivity by improving mass transfer and reactant contact. This is especially advantageous in heterogeneous processes where solid–liquid interactions are essential. Due to the regulated cavitation effects, ultrasound-assisted synthesis produces particles with consistent sizes and shapes (see Chap. 2). Furthermore, ultrasound has improved energy efficiency and process intensification in extraction, crystallization, and emulsification procedures. In biomedical applications, ultrasound's flexibility is further demonstrated by its ability to help with medication delivery and the targeted release of therapeutic substances (Ince 2018). Advanced technologies incorporating other energy sources, such as light or microwaves, further improve ultrasound-assisted techniques. Using the complementary impacts of several energy types, these hybrid systems produce synergistic results that increase product yield and reaction efficiency (Gogate and Pandit 2004a, b).

The special physical phenomena of sonic cavitation, which is essential to sonochemistry, is exploited by ultrasound-assisted methods. High-pressure and low-pressure cycles alternate when ultrasonic vibrations travel through a liquid medium (Doktycz and Suslick 1990). Ultrasound improves mass transfer and allows for more effective reactant mixing, which improves reaction kinetics and product selectivity in chemical synthesis, particularly in heterogeneous systems where distinct phases must interact. Ultrasound, for instance, aids in achieving precise control over particle shape and homogeneous size distribution during the creation of nanoparticles. Additionally, it increases energy efficiency and environmental friendliness (Madhavan et al. 2019).

Ultrasound-assisted drug delivery is also becoming more popular since it facilitates the release of drugs at specific organs, allowing for tailored therapy. Hybrid techniques, which combine ultrasound with other energy sources like light or microwaves, have been investigated in recent breakthroughs. These synergistic systems, called sonophotocatalysis or sonothermal processes, provide enhanced reaction speeds and yields using several energy sources. Sonophotocatalysis, for example, enhances the production of reactive species and improves overall efficiency by combining ultrasonic cavitation with photocatalytic efficiency (Gogate and Pandit 2004a, b). In chemistry, sonochemistry is further subdivided into the following subtypes.

1.4.1 Homogeneous Sonochemistry

It refers to the chemical reactions occurring uniformly throughout the liquid medium due to cavitation. Homogeneous sonochemistry involves the use of ultrasound to improve reactions in single-phase systems. The application of ultrasound generates acoustic cavitation, which produces high pressures and temperatures, accelerating chemical processes (Caulier et al. 1995). While traditionally associated with radical-chain reactions, recent studies show that ultrasound also aids in studying solvation in ionic reactions. In homogeneous systems, ultrasound efficiently supports radical-chain reactions, such as hydrostannation and tin hydride reductions, by providing the energy needed for reaction initiation. (Tuulmets et al. 2010).

Recent developments have investigated hybrid strategies that combine ultrasonic energy with other energy sources, such as microwaves or light. These synergistic systems, sometimes called sonothermal processes or sonophotocatalysis, use several energy sources to

1.4 Ultrasound-Assisted Approaches

enhance reaction speeds and yields. To improve overall efficiency and increase the formation of reactive species, sonophotocatalysis, for example, combines ultrasonic cavitation with photocatalytic activity, which can improve radical-mediated reactions in homogenous systems. For instance, the formation of radicals enhances the efficiency and control of radical polymerization or oxidation processes. Ultrasonic technology makes reactant uniform dispersion possible and also aids in homogenous mixing, which is essential for reaction efficiency (Crum et al. 1999).

Homogeneous sonochemistry can be very helpful in organic synthesis, including the sonochemical oxidation of alcohols and the creation of carbon–carbon bonds (see Chap. 2). Studies have shown that sonication enhances esterification, hydrolysis, and substitution processes by encouraging molecular interactions and enabling the production of intermediate species. Additionally, the approach is used in fields like material sciences and medicines to synthesize molecules with higher yields and purity. Numerous studies have examined the mechanisms behind homogenous sonochemistry, emphasizing the significance of cavitation dynamics. High shear forces and microjets produced by the quick collapse of bubbles improve the overall mechanics. Despite their benefits, research is still being done on the difficulties of scaling up sonochemical processes for commercial use. These restrictions are being addressed by developments in sonochemical reactor design, such as continuous-flow systems and dual-frequency irradiation, allowing sonochemistry to be used more widely across various industries (Luo et al. 2015).

1.4.2 Heterogeneous Sonochemistry

Heterogeneous sonochemistry improves chemical reactions between reactants in various phases, usually in solid–liquid or liquid–liquid systems, known as heterogeneous sonochemistry. Heterogeneous sonochemistry combines ultrasound with other energy sources, such as light, to enhance chemical reactions through synergistic effects. This approach, termed sonophotocatalysis, is an emerging area in process intensification. Ultrasound generates cavitation phenomena, producing localized high-temperature hot spots and intense microjetting that interact with photochemical processes, enhancing reaction efficiency and selectivity. For example, photooxidation of benzyl alcohol, when introducing ultrasound, improved selectivity related to photocatalysis alone. This improvement was attributed to the unique effects of cavitation, such as disrupting the reaction medium and optimizing catalyst behavior (Kegelaers et al. 2000).

This hybrid method, often referred to as a Hybrid Process Intensification Method (HPIM), holds promise for advancing complex reactions and optimizing the photoreactivity of catalysts. Applications include biomass valorization and sustainable chemical transformations, where increased selectivity and efficiency are critical. Developing specialized sonophotoreactors remains a key challenge, but progress in this field could open new avenues for innovative chemical processing in various industries (Chakma and

Moholkar 2015). In heterogeneous systems, the energy generated when these bubbles collapse close to a solid surface or at the interface of immiscible liquids produces intense heat, pressure, and microjets, enhancing mass transfer and producing very reactive species resulting in increased mixing, catalyst activation, and surface cleaning (Ashokkumar 2011).

Ultrasound in catalysis improves phase-to-phase interaction and helps solid catalysts disperse more easily. For instance, sonication can activate or regenerate metal oxides and supported catalysts more effectively. Another well-reported use of ultrasound is in environmental remediation, namely in the degradation of pollutants. It aids in breaking down toxins that have been adsorbed on solid surfaces or suspended in aqueous solutions. Recent developments have concentrated on developing hybrid systems along with ultrasound with additional energy sources, such as light or microwaves, to improve reaction efficiency further. These synergistic effects result from the design of sonophotocatalytic and sonoelectrochemical reactors, which are being investigated for sustainable chemical processes. By utilizing the unique characteristics of sonic cavitation, heterogeneous sonochemistry offers intriguing prospects for cleaner, more effective chemical processes (Safari and Zarnegar 2014).

1.4.3 Dual-Frequency Ultrasound

Dual-frequency sonochemistry introduces a novel approach to enhancing chemical processes by combining ultrasound at two frequencies. This technique optimizes cavitation effects, improving reaction efficiency and scalability. Dual-frequency irradiation has been shown to influence bubble dynamics significantly. These models account for liquid compressibility and explain the enhanced performance of dual-frequency systems related to single-frequency setups. The synergy between frequencies affects cavitation bubbles' intensity and collapse conditions, improving sonochemical yields. Dual-frequency sonochemistry is an advanced approach to ultrasound-assisted chemical processes (Kanthale et al. 2007). Combining two ultrasonic frequencies, either harmonically or non-harmonically, creates enhanced cavitation effects, improving the efficiency and control of chemical reactions. This can lead to more intense energy release and a higher desired reaction yield than traditional single-frequency sonochemistry. Factors such as phase shifts and power distribution between the frequencies are crucial in optimizing the process (Brotchie et al. 2008).

Dual-frequency ultrasonography is a cutting-edge sonochemical technique intended to increase the effectiveness and regulation of chemical processes. In sonochemical processes, cavitation creating and dissolving bubbles in a liquid is crucial. Dual-frequency ultrasound enhances bubble behavior due to increased energy transfer and enhances response results (Tatake and Pandit 2002). Bubble size and collapse dynamics are influenced by the complicated acoustic fields created when two frequencies are combined.

The bubbles' resonance conditions restrict cavitation in conventional single-frequency systems. Dual-frequency ultrasonography can provide a synergistic impact by allowing bubbles to oscillate at one frequency and collapse at another. This results in higher temperatures and pressures inside the bubbles, more intense cavitation events, and a more significant generation of reactive species, including hydroxyl radicals. Additionally, the improved cavitation enhances mixing and mass transfer, both essential for effective chemical reactions (Esclapez et al. 2011).

The capacity to regulate the phase connection between the two frequencies is another benefit of dual-frequency ultrasonography. The cavitation effects may be amplified or attenuated by the phase difference. Researchers have demonstrated that by producing ideal waveforms that promote bubble collapse, likely phase shifts can optimize sonochemical activity. Dual-frequency systems are quite versatile for various reactions due to their ability to change frequencies and phase shifts. (Ashokkumar 2011). Applications of dual-frequency sonochemistry are diverse, ranging from chemical synthesis and degradation reactions to environmental and industrial processes. Additionally, its ability to manipulate bubble coalescence and cavitation behavior in the presence of solutes opens up possibilities for tailoring reactions in more complex systems. With ongoing reactor design and modeling advancements, dual-frequency sonochemistry is promising for scaling up and innovating chemical processing technologies (Son et al. 2010). Despite its advantages, dual-frequency ultrasound application presents difficulties, especially when constructing reactors that can efficiently combine and regulate multiple frequencies. Phase control, power settings, and frequency combinations must be optimized to get the intended results using the dual-frequency systems (Adamou et al. 2024).

1.4.4 Ultrasound in Microreactors

The integration of ultrasound into microreactors has opened new frontiers in chemical engineering and process intensification. Microreactors, known for their high area-to-volume ratio and accurate control over reaction conditions, benefit significantly from ultrasound's ability to induce cavitation. This combination is particularly valuable in homogeneous and heterogeneous reactions where mixing and dispersion are critical (Banakar et al. 2022). The small dimensions of microreactors allow better control over ultrasound intensity, ensuring uniform energy distribution. This reduces side reactions and improves selectivity in chemical transformations. Moreover, the continuous-flow nature of microreactors synergizes with ultrasound to achieve high yields and scalability while maintaining reaction consistency (Wang et al. 2024). Recent advancements in designing ultrasonic microreactors focus on integrating piezoelectric transducers and optimizing flow paths for effective energy coupling. These innovations enable real-time process monitoring and efficient energy usage, making ultrasonic microreactors an eco-friendly and scalable option. This emerging technology demonstrates great potential for transforming conventional batch processes into efficient, sustainable, cost-effective, continuous operations (Adhi et al. 2024).

Microreactor technologies aided by ultrasound have drawn a lot of interest because they can improve chemical reactions by precisely controlling the reaction parameters. The production of hot spots and strong shear forces within the reactor's limited volume as well as acoustic cavitation, is a unique property utilized when integrating ultrasound into microreactors. Microreactors, in contrast to traditional reactors, have a high surface-to-volume ratio, which enhances the effect of ultrasound and enables improved control over heat and mass transmission, reducing the development of hot and cold patches (Gogate and Kabadi 2009). Ultrasound is particularly useful in microreactors for reactions that need fine control over reaction parameters or are diffusion-limited. Higher reaction rates and more uniform reactant distribution are made possible by the vigorous mixing produced by ultrasonic waves, which can also break down diffusion barriers. According to research, ultrasonography can also help microreactors be cleaned and kept from fouling, which would increase their operating efficiency (Sakthipandi et al. 2024). The synthesis of different organic compounds has also made use of ultrasound-driven microreactors, where the combination of precise temperature control and cavitation effects results in shorter reaction times and greater yields. Studies show that by maximizing reaction conditions and reducing waste creation, these hybrid systems can drastically lower energy consumption and environmental impact (Adamou et al. 2024).

Ultrasound plays a vital role in the formation of esters, amides, and other functional groups. It facilitates reactions by breaking intermolecular forces and enhancing the reactivity of substrates. Ultrasound-assisted catalysis improves catalyst dispersion, increases surface area, and accelerates reaction kinetics. Sonochemical routes are also employed to synthesize materials, such as graphene derivatives and metal–organic frameworks (MOFs). Ultrasound-assisted degradation of pollutants, known as sonolysis, is an effective method for water treatment. The intense conditions of cavitation break down complex organic molecules into more straightforward, less harmful substances. Ultrasound is also used in processes like ultrasonic sludge disintegration and soil remediation.

It enhances the extraction of organic as well as inorganic compounds from plants and improves the stability of emulsions. In pharmaceuticals, ultrasound aids in drug formulation and enhances the solubility and bioavailability of poorly soluble drugs (Osman et al. 2024).

1.5 Sonochemical Catalysis

One of the principles of green chemistry, catalysis, based on ultrasound principles, has been widely used in chemistry, medicine, and biology. Numerous synthetic approaches for the synthesis and preparation of nano/microstructures, such as gas phase, liquid phase, and mixed-phase methods, have been developed due to the quick advances in nanoscience and nanotechnology. Many chemical, physical, and mechanical characteristics that frequently dictate the uses of nano-, as well as microstructured devices, can occasionally be highly reliant on the methods used to synthesize materials; hence, a suitable technology will be a driving factor behind the development of novel synthesis processes (Li et al. 2019). It has proven powerful and unique when preparing micro- and nanostructured substances, such as polymers and combination materials. Furthermore, mechanical and physical forces contribute significantly to its creation (Bhosale et al. 2016).

Generally, it is challenging to pinpoint the precise impacts of ultrasonic radiation and to demonstrate how it might alter reactivity. The primary outcomes, however, are intriguing and include increased yield, improved rate or selectivity, and decreased response time. The traditional stirring approach with ultrasonic-assisted catalytic transesterification has recently employed ZnO as a nanocatalyst. The sonochemical term shortened the reaction time from 75 min while maintaining a 96% yield at a fixed temperature (Varghese et al. 2018). Also, the use of Fe_3O_4 nanocatalysts under experimental conditions of 20 kHz ultrasonic frequency (probe), 180 W power, 45 min reaction time, and a temperature of 60 °C has been successfully applied for the production of secondary amides via the Beckmann rearrangement.

1.6 Conclusion

Ultrasound has various applications in the fields of pharmaceutical sciences, engineering, and other fields. It is gaining a wide range of attention from researchers all over the globe. This chapter has highlighted the basics of ultrasound along with its perspective from a chemistry point of view. Further, the chapter highlights the combination of ultrasound with different technologies. Thus, this chapter paves the way for dwelling in depth in the world of ultrasound by giving a preface for the consecutive chapters.

References

Adamou P et al (2024) Ultrasonic reactor set-ups and applications: a review. Ultrason Sonochem 107:106925. https://doi.org/10.1016/j.ultsonch.2024.106925

Adewuyi YG (2005) Sonochemistry in environmental remediation. 2. Heterogeneous sonophotocatalytic oxidation processes for the treatment of pollutants in water. Environ Sci Technol 39(22):8557–8570. https://doi.org/10.1021/es0509127

Adhi TP, Aqsha A, Indarto A (2024) Comparative studies on thermal, microwave-assisted, and ultrasound-promoted preparations. In: Green chemical synthesis with microwaves and ultrasound. Wiley, pp 337–380. https://doi.org/10.1002/9783527844494.ch12

Ashokkumar M (ed) (2011) Theoretical and experimental sonochemistry involving inorganic systems. Springer, Netherlands, Dordrecht. https://doi.org/10.1007/978-90-481-3887-6

Athanassiadis AG et al (2022) Ultrasound-responsive systems as components for smart materials. Chem Rev 122(5):5165–5208. https://doi.org/10.1021/acs.chemrev.1c00622

Banakar VV et al (2022) Ultrasound assisted continuous processing in microreactors with focus on crystallization and chemical synthesis: a critical review. Chem Eng Res des 182:273–289. https://doi.org/10.1016/j.cherd.2022.03.049

Bhargava N, Mor RS, Kumar K, Sharanagat VS (2021) Advances in application of ultrasound in food processing: a review. Ultrason Sonochem 70:105293. https://doi.org/10.1016/j.ultsonch.2020.105293

Bhosale MA, Chenna DR, Bhanage BM (2016) One-step sonochemical irradiation dependent shape controlled crystal growth study of gold nano/microplates with high catalytic activity in degradation of dyes. ChemistrySelect 1(3):504–512. https://doi.org/10.1002/slct.201600044

Brotchie A, Grieser F, Ashokkumar M (2008) Sonochemistry and sonoluminescence under dual-frequency ultrasound irradiation in the presence of water-soluble solutes. J Phys Chem C 112(27):10247–10250. https://doi.org/10.1021/jp801763v

Carlton JS (2019) Cavitation. In: Marine propellers and propulsion. Elsevier, pp 217–260. https://doi.org/10.1016/B978-0-08-100366-4.00009-2

Caulier TP, Maeck M, Reisse J (1995) Homogeneous sonochemistry: a study of the induced isomerization of 1,2-dichloroethene under ultrasonic irradiation. J Org Chem 60(1):272–273. https://doi.org/10.1021/jo00106a049

Chakma S, Moholkar VS (2015) Intensification of wastewater treatment using sono-hybrid processes: an overview of mechanistic synergism. Indian Chem Eng 57(3–4):359–381. https://doi.org/10.1080/00194506.2015.1026948

Crum LA et al (eds) (1999) Sonochemistry and sonoluminescence. Springer, Netherlands, Dordrecht. https://doi.org/10.1007/978-94-015-9215-4

Dehghani MH, Najafpoor AA, Azam K (2010) Using sonochemical reactor for degradation of LAS from effluent of wastewater treatment plant. Desalination 250(1):82–86. https://doi.org/10.1016/j.desal.2009.05.011

Doktycz SJ, Suslick KS (1990) Interparticle collisions driven by ultrasound. Science 247(4946):1067–1069. https://doi.org/10.1126/science.2309118

Duck FA, Thomas AMK (2022) Paul Langevin (1872–1946): the father of ultrasonics

Draye M, Kardos N (2016) Advances in green organic sonochemistry. Top Curr Chem 374(5):74. https://doi.org/10.1007/s41061-016-0074-7

Esclapez MD et al (2011) Ultrasound-assisted extraction of natural products. Food Eng Rev 3(2):108–120. https://doi.org/10.1007/s12393-011-9036-6

Gadge SS et al (2024) Enhanced sunlight-driven catalysis for hydrogen generation and dye remediation using synergistic p-Co_3O_4/n-TiO_2 nanocomposites. Nanoscale Adv 6(6):1661–1677. https://doi.org/10.1039/D3NA01167D

Gogate PR, Kabadi AM (2009) A review of applications of cavitation in biochemical engineering/biotechnology. Biochem Eng J 44(1):60–72. https://doi.org/10.1016/j.bej.2008.10.006

Gogate PR, Pandit AB (2004a) A review of imperative technologies for wastewater treatment I: oxidation technologies at ambient conditions. Adv Environ Res 8(3–4):501–551. https://doi.org/10.1016/S1093-0191(03)00032-7

Gogate PR, Pandit AB (2004b) Sonophotocatalytic reactors for wastewater treatment: a critical review. AIChE J 50(5):1051–1079. https://doi.org/10.1002/aic.10079

Guo C, Zhu X (2018) Effect of ultrasound on dynamics characteristic of the cavitation bubble in grinding fluids during honing process. Ultrasonics 84:13–24. https://doi.org/10.1016/j.ultras.2017.09.016

References

Hoo DY et al (2022) Ultrasonic cavitation: an effective cleaner and greener intensification technology in the extraction and surface modification of nanocellulose. Ultrason Sonochem 90:106176. https://doi.org/10.1016/j.ultsonch.2022.106176

Humphrey VF (2007) Ultrasound and matter—physical interactions. Progr Biophys Molec Biol 93:195–211. https://doi.org/10.1016/j.pbiomolbio.2006.07.024

Ince NH (2018) Ultrasound-assisted advanced oxidation processes for water decontamination. Ultrason Sonochem 40:97–103. https://doi.org/10.1016/j.ultsonch.2017.04.009

Kanthale PM, Gogate PR, Pandit AB (2007) Modeling aspects of dual frequency sonochemical reactors. Chem Eng J 127(1–3):71–79. https://doi.org/10.1016/j.cej.2006.09.023

Kasprzyk-Hordern B, Ziółek M, Nawrocki J (2003) Catalytic ozonation and methods of enhancing molecular ozone reactions in water treatment. Appl Catal B 46(4):639–669. https://doi.org/10.1016/S0926-3373(03)00326-6

Kegelaers Y, Delplancke J-L, Reisse J (2000) Sonochemistry: scope, limitations… and artifacts. Chimia 54(1–2):48. https://doi.org/10.2533/chimia.2000.48

Leighton TG, Apfel RE (1994) The acoustic bubble. J Acous Soc Am 96(4):2616–2616. https://doi.org/10.1121/1.410082

Leong T, Ashokkumar M, Kentish S (2016) The growth of bubbles in an acoustic field by rectified diffusion. In: Handbook of ultrasonics and sonochemistry. Springer Singapore, Singapore, pp 69–98. https://doi.org/10.1007/978-981-287-278-4_74

Li Z et al (2019) Ultrasonic-assisted fabrication and release kinetics of two model redox-responsive magnetic microcapsules for hydrophobic drug delivery. Ultrason Sonochem 57:223–232. https://doi.org/10.1016/j.ultsonch.2019.04.037

Li Z et al (2021) Ultrasonic cavitation at liquid/solid interface in a thin Ga–In liquid layer with free surface. Ultrason Sonochem 71:105356. https://doi.org/10.1016/j.ultsonch.2020.105356

Luo J et al (2015) Fundamentals of acoustic cavitation in sonochemistry, pp 3–33. https://doi.org/10.1007/978-94-017-9624-8_1

Madhavan J et al (2019) Hybrid advanced oxidation processes involving ultrasound: an overview. Molecules 24(18):3341. https://doi.org/10.3390/molecules24183341

Mason TJ (2009) Sonochemistry—beyond synthesis [WWW Document]. RSC Education. https://edu.rsc.org/feature/sonochemistry-beyond-synthesis/2020230.article (accessed 11.21.24)

Margulis et al (1985) Sonoluminescence and sonochemical reactions in cavitation fields. A review. Ultrasonics 23(4):157–169. https://doi.org/10.1016/0041-624X(85)90024-12.

Mason TJ, Lorimer JP (2002) Applied sonochemistry. Wiley. https://doi.org/10.1002/352760054X

Mehta N et al (2022) Ultrasound-assisted extraction and the encapsulation of bioactive components for food applications. Foods 11(19):2973. https://doi.org/10.3390/foods11192973

Meng X, Zhang Z, Li X (2015) Synergetic photoelectrocatalytic reactors for environmental remediation: a review. J Photochem Photobiol, C 24:83–101. https://doi.org/10.1016/j.jphotochemrev.2015.07.003

Mohamed et al (2016) Sonochemistry (Applications of ultrasound in chemical synthesis and reactions): A review part I. Al-Azhar Journal of Pharmaceutical Sciences 53(1) 108–122. https://doi.org/10.21608/ajps.2016.6890

Neppiras EA (1980) Acoustic cavitation. Phys Rep 61(3):159–251. https://doi.org/10.1016/0370-1573(80)90115-5

Neppolian B et al (2012) Graphene oxide based Pt–TiO_2 photocatalyst: ultrasound assisted synthesis, characterization and catalytic efficiency. Ultrason Sonochem 19(1):9–15. https://doi.org/10.1016/j.ultsonch.2011.05.018

Newman PG, Rozycki GS (1998) The history of ultrasound. Surg Clin North Am 78:179–195. https://doi.org/10.1016/S0039-6109(05)70308-X

Osman AI et al (2024) Coordination-driven innovations in low-energy catalytic processes: advancing sustainability in chemical production. Coord Chem Rev 514:215900. https://doi.org/10.1016/j.ccr.2024.215900

Pokhrel N, Vabbina PK, Pala N (2016) Sonochemistry: science and engineering. Ultrason Sonochem 29:104–128. https://doi.org/10.1016/j.ultsonch.2015.07.023

Rosales Pérez A, Esquivel Escalante K (2024) The evolution of sonochemistry: from the beginnings to novel applications. ChemPlusChem 89(6). https://doi.org/10.1002/cplu.202300660

Safari J, Zarnegar Z (2014) Ultrasonic activated efficient synthesis of chromenes using aminosilane modified Fe_3O_4 nanoparticles: a versatile integration of high catalytic activity and facile recovery. J Mol Struct 1072:53–60. https://doi.org/10.1016/j.molstruc.2014.04.023

Safwa SM, Ahmed T, Talukder S, Sarker A, Rana MR (2023) Applications of non-thermal technologies in food processing industries-a review. J Agric Food Res:100917. https://doi.org/10.1016/j.jafr.2023.100917

Sakthipandi K et al (2024) Ultrasound-based sonochemical synthesis of nanomaterials. In: Handbook of vibroacoustics, noise and harshness. Springer Nature Singapore, Singapore, pp 1–46. https://doi.org/10.1007/978-981-99-4638-9_58-1

Singla M, Sit N (2021) Application of ultrasound in combination with other technologies in food processing: a review. Ultrason Sonochem 73:105506. https://doi.org/10.1016/j.ultsonch.2021.105506

Son Y et al (2010) Estimation of sonochemical reactions under single and dual frequencies based on energy analysis. Japanese J Appl Phys 49(7S):07HE02. https://doi.org/10.1143/JJAP.49.07HE02

Suslick KS (1990) Sonochemistry. Science 247(4949):1439–1445. https://doi.org/10.1126/science.247.4949.1439

Suslick KS (2003) 1.41—Sonochemistry. In: McCleverty JA, Meyer TJ (eds) Comprehensive coordination chemistry II. Pergamon, Oxford, pp 731–739. https://doi.org/10.1016/B0-08-043748-6/01046-X

Tatake PA, Pandit AB (2002) Modelling and experimental investigation into cavity dynamics and cavitational yield: influence of dual frequency ultrasound sources. Chem Eng Sci 57(22–23):4987–4995. https://doi.org/10.1016/S0009-2509(02)00271-3

Thompson LH, Doraiswamy LK (1999) Sonochemistry: science and engineering. Ind Eng Chem Res 38(4):1215–1249. https://doi.org/10.1021/ie9804172

Tuulmets A et al (2010) Sonochemistry of homogeneous ionic reactions. Mini-Rev Org Chem 7(3):204–211. https://doi.org/10.2174/157019310791384155

Varghese R, Henry JP, Irudayaraj J (2018) Ultrasonication-assisted transesterification for biodiesel production by using heterogeneous ZnO nanocatalyst. Environ Prog Sustainable Energy 37(3):1176–1182. https://doi.org/10.1002/ep.12770

Wang S et al (2024) Design and optimization of novel vortex microreactors for ultrasound-assisted synthesis of high-performance Fe_3O_4 nanoparticles. Chem Eng J 501:157672. https://doi.org/10.1016/j.cej.2024.157672

Wu Z et al (2020) Sonozonation (sonication/ozonation) for the degradation of organic contaminants—a review. Ultrason Sonochem 68:105195. https://doi.org/10.1016/j.ultsonch.2020.105195

Yao Y, Pan Y, Liu S (2020) Power ultrasound and its applications: a state-of-the-art review. Ultrason Sonochem 62:104722. https://doi.org/10.1016/j.ultsonch.2019.104722

Yasui K (2022) Production of O radicals from cavitation bubbles under ultrasound. Molecules 27(15):4788. https://doi.org/10.3390/molecules27154788

Yusof NSM et al (2016) Physical and chemical effects of acoustic cavitation in selected ultrasonic cleaning applications. Ultrason Sonochem 29:568–576. https://doi.org/10.1016/j.ultsonch.2015.06.013

References

Zigangareeva LM, Klselev OM (1998) Subsonic compressible cavitation flow past a plate at small cavitation numbers. Fluid dynamics 33(4):543–551.

Zupanc M et al (2019) Effects of cavitation on different microorganisms: the current understanding of the mechanisms taking place behind the phenomenon. A review and proposals for further research. Ultrason Sonochem 57:147–165. https://doi.org/10.1016/j.ultsonch.2019.05.009

Applications in Chemistry

Abstract

Sonochemistry has emerged as a promising technology for the synthesis of organic compounds along with inorganic nanomaterials. It promoted the reaction in homogenous and heterogeneous mediums with excellent product outcomes and remarkable selectivities. This method catalyzes the reactions in the absence of the catalyst through the cavitation effects via the formation of ionic or radical intermediates as primary effects, and mechanical effects resulting from the cavitation are termed secondary effects. This book chapter discusses the various types of sonochemical reactions, mechanisms and reactivity of ultrasound irradiation and its applications in nanomaterial preparations. The advantages of ultrasound-assisted synthesis compared to conventional methods were critically analyzed.

2.1 Introduction

Immediately after a decade of discovery, sonochemistry emerged as a promising and cost-effective approach for sustainable chemical synthesis, addressing environmental concerns associated with traditional chemical processes (Draye et al. 2019). Even today, it has the potential to significantly enhance our efforts to develop more environmentally friendly chemical processes. This affordable approach provides synthetic chemists with a valuable tool to activate molecules, opening up opportunities for various applications in chemical transformations (Chatel and Colmenares 2017). As discussed in Chap. 1, cavitation plays a crucial role in sonochemistry, making it essential for at least one phase of the reaction mixture to be liquid to facilitate effective sonochemical reactions. Furthermore, reactive species such as radicals, ions, and excited molecules can also be generated via hot spots. Those species are indispensable in accelerating the rate of slow chemical reactions that do not proceed with conventional approaches (Rosales Pérez and Esquivel Escalante 2024). The use of milder reaction conditions with improved energy efficiency makes this approach an excellent choice for exploring innovative pathways in green chemistry, contributing to sustainability and opening the door to developing more effective and environmentally friendly chemical processes (Cintas and Luche 1999; Machado et al. 2021). As a result, sonochemistry emerges as a valuable tool in chemical reactions, offering exciting opportunities to enhance and modify established chemical processes. Moreover, the expanding applications of sonochemistry in synthesis provide worthwhile opportunities for chemists and stimulate greater interest in industry and chemical engineering (Martínez et al. 2021). This emerging area encourages collaboration and innovation, paving the way for new advancements in the fields of pharmaceuticals, material science, agriculture, and wastewater treatment (Pirsaheb and Moradi 2020; Bao et al. 2023; Pirsaheb et al. 2023).

The present chapter focuses on the applications of ultrasounds in catalyzing organic reactions and positions it as a greener approach compared to conventional technologies. In addition, the utilization of this approach will be positively examined for its potential applications in material sciences and the synthesis of nanomaterials, highlighting opportunities for advancements in these fields. The underlying mechanism and effects of ultrasound on the reactivities have been thoroughly discussed.

2.2 Ultrasound in Synthetic Organic Chemistry

Synthetic organic chemistry played a vital role in the advancements of the twentieth century, which greatly enhanced public health through the discovery and development of life-saving drugs, perfumes and high-quality textiles. This field has transformed various sectors and improved the quality of life for many. Various advanced technologies have been utilized to process synthetic organic materials, and a few of them are chemistry via the use of microwaves, electrochemistry, photochemistry, flow chemistry and sonochemistry. Like another process, sonochemistry also activates the formation and dissociation of chemical bonds (Li et al. 2021). The activation occurs when ultrasonic waves above 20 kHz are emitted in a liquid, weakening and breaking chemical bonds (sonolysis), which induces the formation of free radicals and ultimately promotes the synthetic reaction. When water serves as the reaction medium, it leads to the generation of a range of oxidizing and reducing products through dissociation into hydrogen and hydroxyl free radicals, highlighting the crucial role of water in facilitating diverse chemical reactions (Machado et al. 2021). Initially, research in sonochemistry appeared primarily as a method for initiating the reaction that needed activation, particularly for those dependent on activating metallic or solid reagents. However, the development over the decade has revealed much broader applicability than previously thought. Hence, it is paramount to investigate the sonochemical effects to expedite the organic reactions and involved mechanistic

2.2 Ultrasound in Synthetic Organic Chemistry

pathways. The organic reactions are categorized into three different types: homogeneous sonochemistry, heterogeneous sonochemistry, and sonocatalytic reaction.

2.2.1 Homogeneous Sonochemistry

In homogeneous sonochemical reactions, the liquid medium generates the radicals via cavitation. Most commonly, ultrasound lysis of water medium leads to H and OH radical formation, which is further converted into H_2 and H_2O_2, acting as oxidizing and reducing agents. This product formation speeds up the oxidation and reduction reactions. The ultrasound waves were also used in various reactions, including Knoevenagel condensation, the Biginelli reaction, benzoin condensation, the Hantzsch reaction, etc.

In this setting, Rashid et al. have employed sodium bicarbonate for the synthesis of the products of the Biginelli reaction, i.e., 2-amino-1,4-dihydropyrimidines in approximately 81% yield from guanidine hydrochloride, 1,3-dicarbonyl compounds, and substituted aldehydes in DMF using ultrasonic irradiation (Scheme 2.1). Significantly, the reaction was completed in just 55 min under ultrasound irradiation, while in the absence of ultrasound, it yielded only 27% and required approximately 25 h to complete. Ultrasonic waves play a crucial role in catalyzing reactions, reducing reaction time and enhancing overall yield (Ahmad et al. 2016).

In another study, Yang et al. synthesized the various thioethers from aryl thiophenols, substituted benzaldehydes and dimedone in shorter reaction time with enhanced output of products under ultrasound using *p*-dodecylbenzene sulfonic acid as a catalyst (Scheme 2.2). The present method is environmentally friendly, has an easy isolation protocol for the product, and is a novel alternative to synthesizing thioether (Ahmad et al. 2016).

The oxidation of substituted primary alcohols to their respective aldehydes was carried out by Naik et al. using nitric acid and ferrous chloride (Scheme 2.3). The chemoselective approach under ultrasound provides one product, i.e., aldehydes, over the other, i.e., acids.

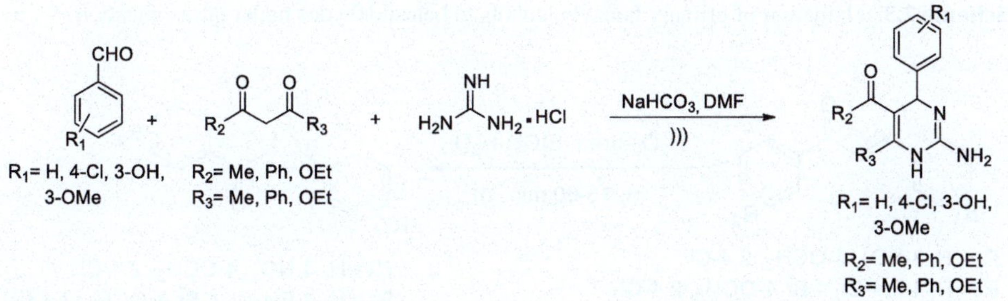

Scheme 2.1 Sonochemical synthesis of 2-amino-1, 4-dihydropyrimidines via Hantzsch reaction

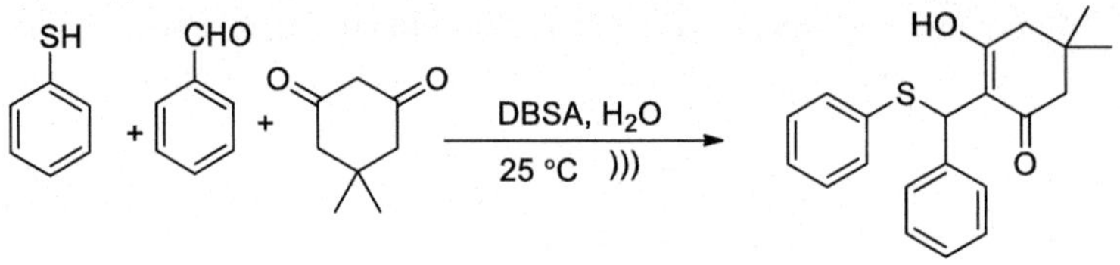
SH
CHO
DBSA, H2O
25 °C)))
HO
S
O

Scheme 2.2 Synthesis of thioethers using ultrasonic irradiations

The reaction proceeded in a 35 kHz ultrasonic bath and yielded substituted benzaldehydes at room temperature within 10 to 25 min without impurity (Song et al. 2015).

Jin et al. studied the epoxidation reaction of chalcones in the presence of H_2O_2-urea as an eco-friendly oxidizing agent with ultrasonic waves (Scheme 2.4). The present methodology presents several advantageous features, including the use of milder reaction conditions that minimize the risk of unwanted side reactions. It employs safer solvents, significantly reducing potential hazards associated with more volatile or toxic alternatives. Additionally, this approach shortens the overall reaction time, enhancing efficiency in the process. Furthermore, it utilizes a comparatively safer oxidizing system, which improves safety and contributes to a more environmentally friendly practice (Jin et al. 2009).

Aryl carbon–carbon coupling reaction between the 4-chlorothieno[2,3-d]pyrimidines with substituted *N*-substituted indole to synthesize 4-(*N*-substitutedindol-3-yl)thieno[2,3-*d*]pyrimidines which is accomplished in the presence of ultrasound using acetic acid as promotor (Scheme 2.5). This sonochemical heteroarylation method provides several advantages: a simple reaction protocol, reduced reaction time, utilization of an eco-friendly energy source, and elimination of the need for metal catalysts. Furthermore, the

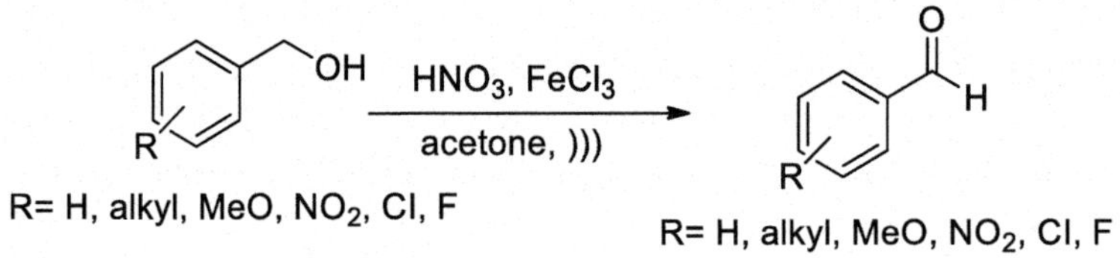
OH
HNO3, FeCl3
acetone,)))
R
O
H
R= H, alkyl, MeO, NO2, Cl, F
R= H, alkyl, MeO, NO2, Cl, F

Scheme 2.3 Oxidation of primary benzyl alcohols to benzaldehydes under ultrasonication

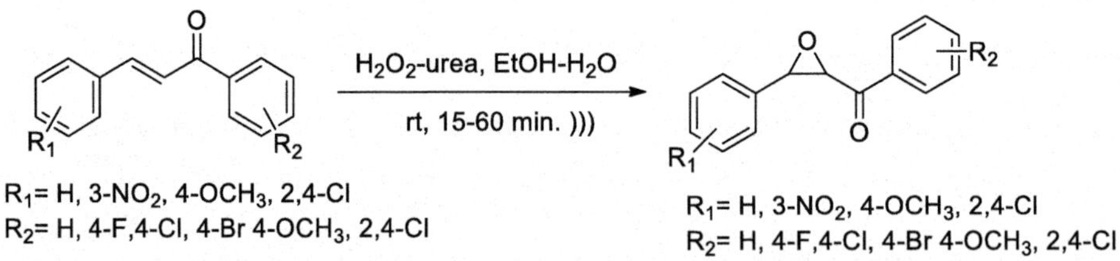
O
H2O2-urea, EtOH-H2O
rt, 15-60 min.)))
R1
R2
R1= H, 3-NO2, 4-OCH3, 2,4-Cl
R2= H, 4-F,4-Cl, 4-Br 4-OCH3, 2,4-Cl
R1= H, 3-NO2, 4-OCH3, 2,4-Cl
R2= H, 4-F,4-Cl, 4-Br 4-OCH3, 2,4-Cl

Scheme 2.4 Sonochemical epoxidation using urea-hydrogen peroxide

2.2 Ultrasound in Synthetic Organic Chemistry

compounds have also been evaluated in vitro as inhibitors of Tumor necrosis factor alpha (TNF-α) (Adapa et al. 2023).

In another study, Chaudhari et al. developed a mild and convenient one-pot protocol to synthesize 2-aminoselenopyridines from aldehydes, malononitriles, and benzeneselenol in PEG-400 via ultrasonic irradiation (Scheme 2.6) (Chatel and Colmenares 2017).

Aqueous NaCl catalyzed pyrazolo-[3,4-*b*]pyridine derivatives were synthesized via a three-component reaction from aldehydes, ethyl cyanoacetate, and 3-amino-5-methylpyrazole under ultrasound irradiation (Scheme 2.7). The method demonstrates exceptional chemoselectivity for pyrazolo-[3,4-*b*]pyridine over other dehydrogenated compounds. This sonochemical approach offers several advantages, such as an environment-friendly catalyst, simple experiment and workup protocol, and high purity of desired products (Dandia et al. 2014).

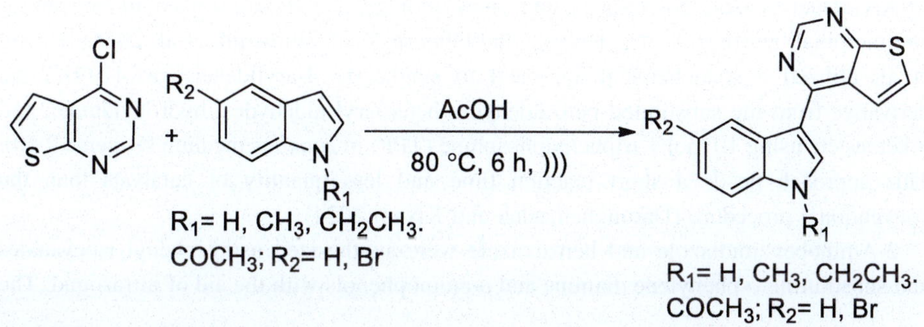

Scheme 2.5 Aryl carbon–carbon-bond formation in homogenous medium via sonications

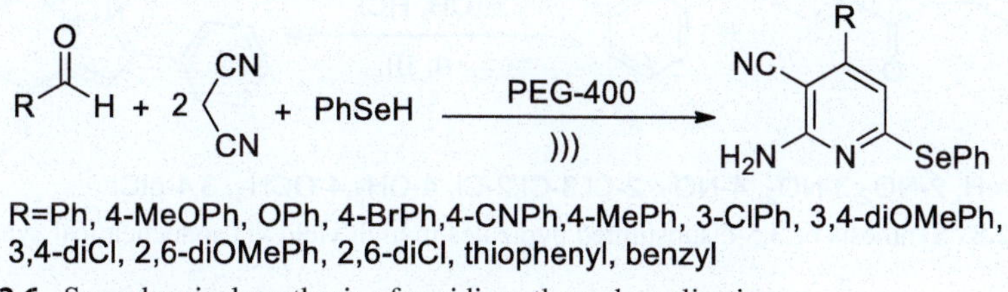

Scheme 2.6 Sonochemical synthesis of pyridines through cyclization

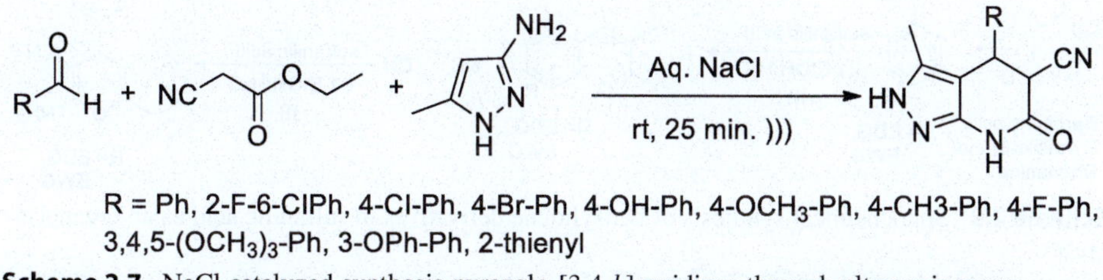

R=Ph, 4-MeOPh, OPh, 4-BrPh, 4-CNPh, 4-MePh, 3-ClPh, 3,4-diOMePh, 3,4-diCl, 2,6-diOMePh, 2,6-diCl, thiophenyl, benzyl

R = Ph, 2-F-6-ClPh, 4-Cl-Ph, 4-Br-Ph, 4-OH-Ph, 4-OCH$_3$-Ph, 4-CH3-Ph, 4-F-Ph, 3,4,5-(OCH$_3$)$_3$-Ph, 3-OPh-Ph, 2-thienyl

Scheme 2.7 NaCl catalyzed synthesis pyrazolo-[3,4-*b*]pyridines through ultrasonic waves

The high-yielding and enviro-friendly protocol was developed via sonochemical synthesis of 5-aryl-1,3-diphenylpyrazole from the condensation reaction of epoxy chalcones and phenylhydrazine at room temperature (Scheme 2.8). This practical protocol offers advantages such as efficient and operationally simple synthesis of desired pyrazoles (Dandia et al. 2014).

Patel et al. synthesized benzylidene derivatives from barbituric acid, pyrazole-5-one, and rhodanine under ultrasonic irradiation using sulfanilic acid as an organocatalyst (Scheme 2.9). This protocol tolerates a wide range of functional groups with an excellent yield of desired products. In addition, this is a mild, green, and environment-friendly procedure to produce a range of benzylidene derivatives from barbituric acid, pyrazole-5-one, and rhodanine from electron-donating and withdrawing aldehydes and further with malononitrile. Those compounds can be used in various industrial applications in the near future (Patel et al. 2024).

Arylideneisoxazole-5(4H)-ones that show biological activities linked to anticancer, antimicrobial, antifungal, etc. (Hatvate and Ghodse 2020; Bhowmik et al. 2024). Kiyani et al. utilized a sonochemical approach to synthesize 4-arylideneisoxazol-5(4H)-one derivative from the substituted benzaldehydes/heteroaryl aldehydes, hydroxylamine, and β-ketoesters using 10 mol% triphenylphosphine (TPP) in aqueous medium (Scheme 2.10). This approach needs a short reaction time and less quantity of catalysts than the conventional procedure (Daroughehzadeh and Kiyani 2024).

2-Aminobenzimdiazole and benzoxazole were synthesized using phenyl isocyanates and substituting o-phenylene diamine and o-aminophenols with the aid of ultrasound. The

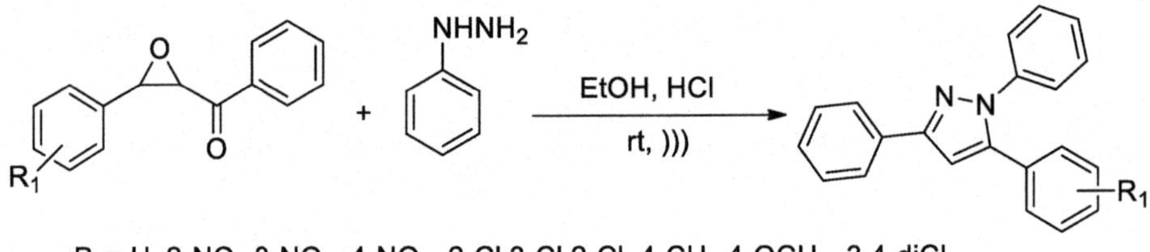

R_1= H, 2-NO$_2$,3-NO$_2$, 4-NO$_2$, 2-Cl,3-Cl,2-Cl, 4-CH$_3$,4-OCH$_3$, 3,4-diCl

Scheme 2.8 Synthesis of 3,5-disubstituted pyrazoles in high yield via sonochemical synthesis

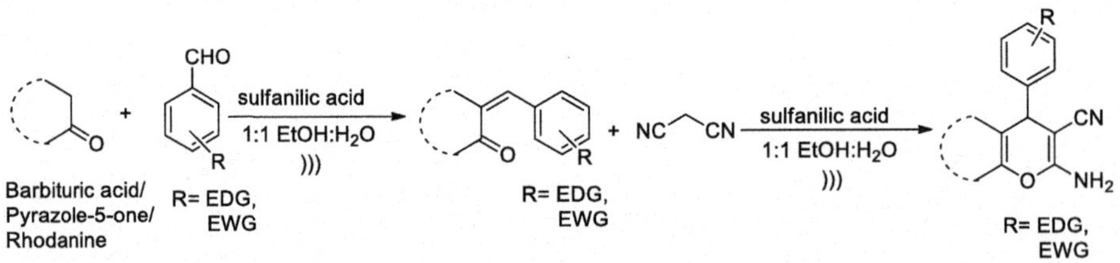

Scheme 2.9 Sonochemical synthesis of benzylidene derivatives in sulfanilic acid as an organocatalyst

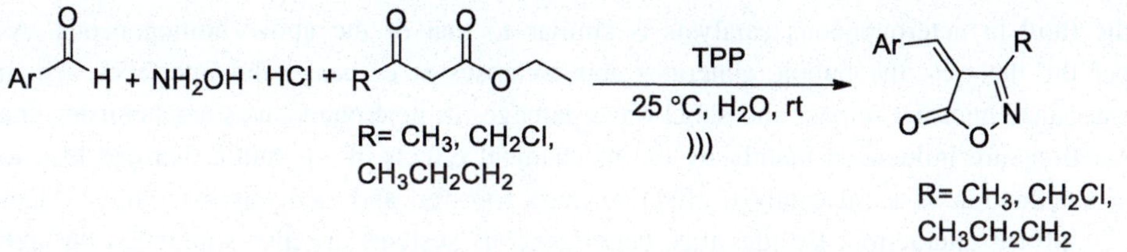

Scheme 2.10 Ultrasound-assisted synthesis of 4-arylideneisoxazol-5(4H)-one derivatives

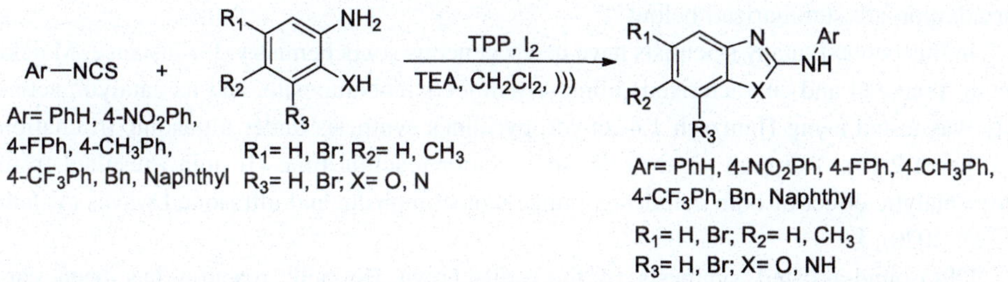

Scheme 2.11 Preparation of 2-aminobenzimdiazole and benzoxazole in TPP using ultrasounds

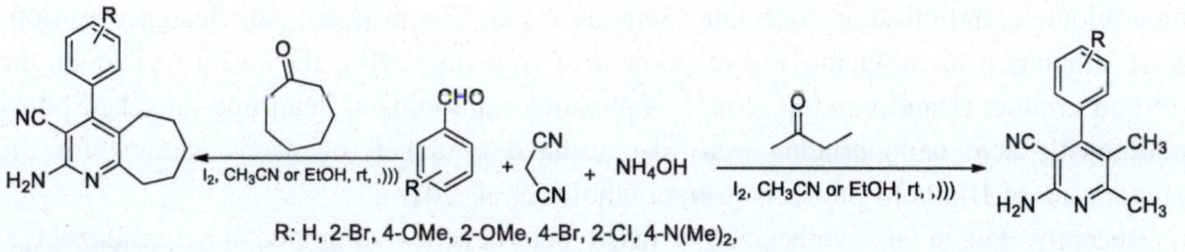

Scheme 2.12 Ultrasound-assisted one-pot synthesis of substituted 2-aminopyrdine-3-nitriles

triphenylphosphine and iodine catalyst systems were utilized along with triethylamine (TEA) for the desulfurization agent (Scheme 2.11). This mild, green, and efficient protocol yielded a high amount of 2-amino benzimidazoles and benzoxazoles.

The efficient synthesis of pyridines was achieved from aliphatic or aryl ketones, substituted benzaldehydes, malononitrile, and a source of nitrogen, i.e., ammonium hydroxide using iodine as a promoter in acetonitrile or ethanol with the assistance of ultrasound (Scheme 2.12). This method is eco-friendly, cost-effective and has an easy workup procedure for separating final products. Furthermore, the compounds were evaluated in silico for glutamine synthetase inhibitory activity (Pagadala et al. 2020).

2.2.2 Heterogeneous Sonochemistry

In organic synthesis, heterogeneous sonochemical reactions involve the reaction between biphasic systems, such as solid–liquid, liquid–liquid, and liquid–gas. The cavitation in the fluid in heterogeneous catalysis is similar to that of the above homogeneous. As per the theories, the bubble generates and collapses at or near solid interfaces, which can cause microjet impact and shockwave damage. In heterogeneous sonochemistry, the reactions are influenced mainly by the mechanical effects of cavitation that can lead to size reduction of solid catalyst, effective mass transfer, and catalysis over the surfaces via surface interaction. Besides this, heterogeneous systems are also stimulated through radical or ionic intermediates generated via cavitational agitation, which we call "false" sonochemistry. Various heterogeneous reactions have been carried out using ultrasonic irradiation are summarized below.

In this context, many scientists have heterogeneous sonochemistry; for instance, Maleki et al. prepared and characterized cellulose/pumice nanocomposite, and its catalytic activity was tested using Hantzsch 1,4-dihydropyridines synthesis under ultrasonic irradiation (Scheme 2.13). The high yield (97%) and shorter reaction time (10 min.) resulted from the catalytic effect of both cellulose/pumice nanocomposite and ultrasound waves (Valadi et al. 2020).

Ultrasound-assisted synthesis of the well-known Biginelli reaction has been carried out by using polyindole under milder reaction conditions. Various bioactive 3,4-dihydropyrimidin-2(1H)-one/thione (DHPM) derivatives were prepared from aldehydes, urea/thiourea, and ethyl acetoacetate (Scheme 2.14). The protocol was designed to optimize efficiency by reducing the reaction time and improving the overall yield of the desired product (Handore et al. 2021). Nanosilica immobilized dendrimer-attached phosphotungstic acid nanoparticles were also utilized as a heterogeneous catalyst for the preparation of Biginelli products (Safaei-Ghomi et al. 2018).

Recently, Pal et al. synthesized a fused heterocyclic compound, i.e., imidazo[1,2-a]pyridin-3-amine from 2-aminopyridines, aldehydes, and isocyanides using Wang-OSO$_3$H as a catalyst in water under ultrasound irradiation (Scheme 2.15). This method is green, eco-friendly and has an excellent yield of desired imidazo[1,2-a]pyridin-3-amine. It is noteworthy to mention that this compound shows remarkable TNF-α inhibitory activity (Shivanoori et al. 2024).

The synthesis of dihydroquinazolinones has also been achieved in heterogeneous system using ultrasound as a promoter. The multicomponent reaction (MCR) of aldehydes, isatoic anhydride, and sources of nitrogen, such as ammonium acetate or aniline, gives

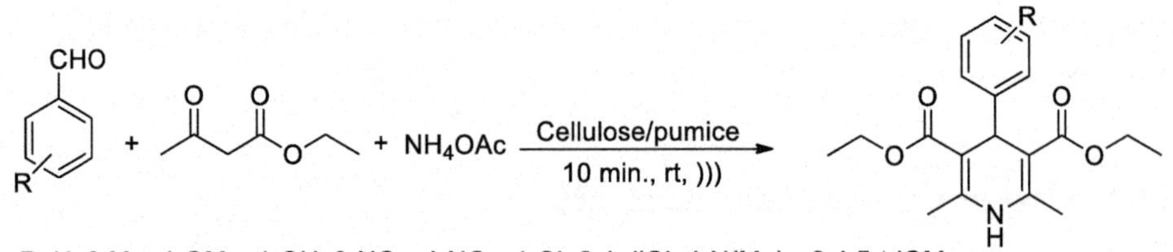

R: H, 3-Me, 4-OMe, 4-OH, 3-NO$_2$, 4-NO$_2$, 4-Cl, 2,4-diCl, 4-N(Me)$_2$, 3,4,5-triOMe,

Scheme 2.13 Sonochemical synthesis of Hantzsch 1,4-dihydropyridines

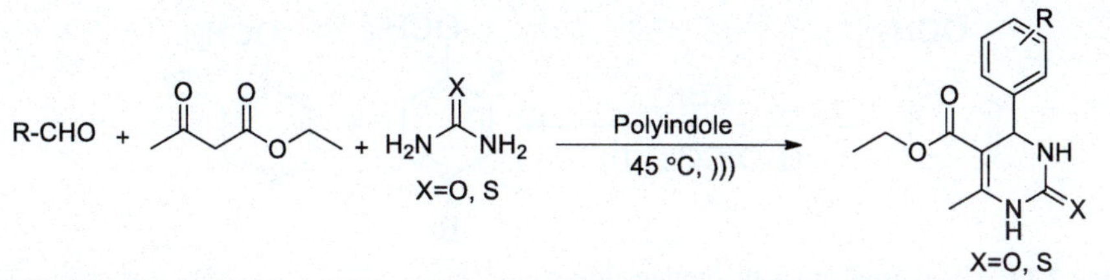

2.2 Ultrasound in Synthetic Organic Chemistry

Scheme 2.14 Synthesis of 4-dihydropyrimidin-2(1H)-one/thione via Biginelli reaction using ultrasonic irradiations

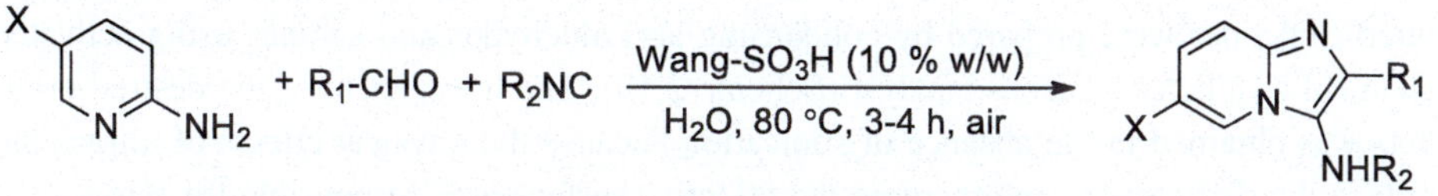

R: Ph, 3-MePh, 4-OMePh, 4-OHPh, 3-OH-4-OMePh, 4-NO_2Ph, 4-ClPh, 2,4-diClPh, 4-N(Me)$_2$Ph, 3,4,5-triOMePh, Thienyl, Cinnamyl

Scheme 2.15 Ultrasound-assisted synthesis of imidazo[1,2-a]pyridin-3-amines

X= H, Cl, CH_3; R_1= Ph, 4-OCH_3Ph, 4-CH_3Ph, 4-ClPh, 4-BrPh, 4-NO_2Ph, cyclohexyl, isobutyl, n-propyl, R_2=t-butyl, benzyl, n-butyl, cyclohexyl

2,3-dihydroquinazolin-4(1H)-ones with the help of ultrasound irradiation. (Scheme 2.16). This synthesis has been catalyzed by the hybrid catalytic system based on carbon nanotube (CNTs) impregnated with metallic nanoparticles (Safari and Gandomi-Ravandi 2017) or multiwallet carbon nanotubes matrices (MWCNTs) (Safari and Gandomi-Ravandi 2014). Those novel methods have many merits, including enhanced yield, convenient workup protocol, recoverable catalyst, and greener synthesis.

Ultrasound-assisted brominations have been carried out by Fujita et al. in heterogeneous liquid–liquid reactions. The anisole has been utilized as a mode substrate for the bromination to convert 4-bromoanisole/2-bromoanisole using a KBr cost-effective brominating agent (Scheme 2.17). Further, the protocol has been applied for the chlorinations as well as bromination. This bromination method is green and simple. The quantity of CCl_4 decides the regioselectivity of the bromination. When the CCl_4 is 10.3 mol eq., it produces 99:3, 4-bromoanisole;2-bromoanisole, respectively (Fujita et al. 2015).

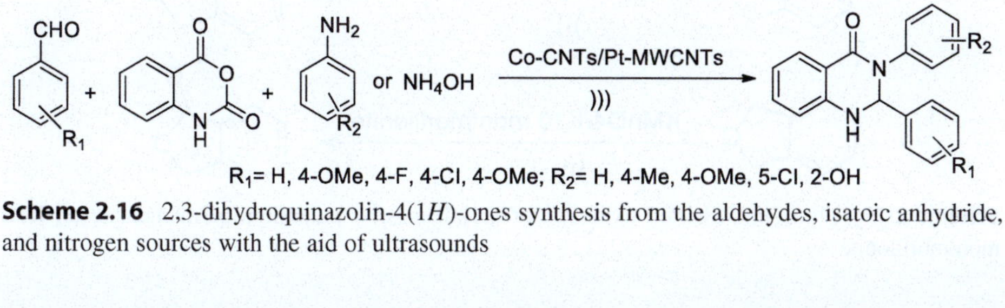

R_1= H, 4-OMe, 4-F, 4-Cl, 4-OMe; R_2= H, 4-Me, 4-OMe, 5-Cl, 2-OH

Scheme 2.16 2,3-dihydroquinazolin-4(1H)-ones synthesis from the aldehydes, isatoic anhydride, and nitrogen sources with the aid of ultrasounds

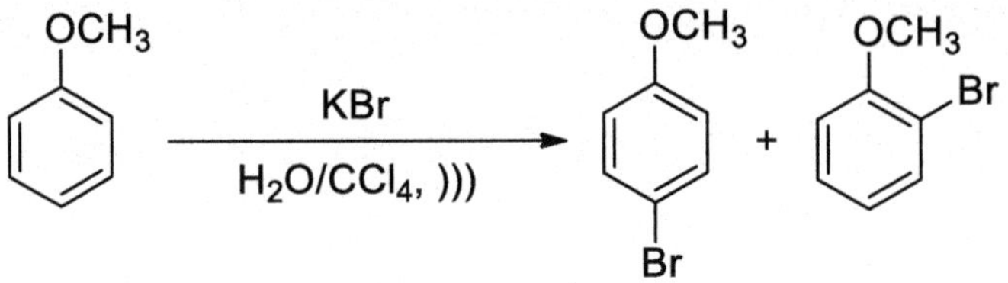

Scheme 2.17 US-assisted aromatic bromination

Condensation reactions are paramount in various industries, such as chemicals, pharmaceuticals, and polymers, with an impressive atom economy. The ultrasounds promoted various such reactions; one of these reactions is Claisen–Schmidt. The various substituted chalcones were prepared by condensing aryl aldehydes and ketones using activated charcoal as a heterogeneous catalyst (Scheme 2.18). The lower yield of the desired products was obtained in the absence of sonication because the physical effects of ultrasound reduce the particle size and increase the effective surface area responsible for this activation. The synthesis chalcones were tested as an antibacterial, and all the compounds have shown excellent activity (Calvino et al. 2006).

Chemoselective oxidation of alkylarenes was achieved by Shaabani et al. using a potassium permanganate on K10 montmorillonite under ultrasonic irradiation (Scheme 2.19). The ultrasonic irradiation improves the overall yield of the above reaction (Shaabani et al. 2002).

Not only an oxidation but also a growing body of research highlights the potential of ultrasounds in the reduction reactions. In a similar context, zinc reduced the benzophenone

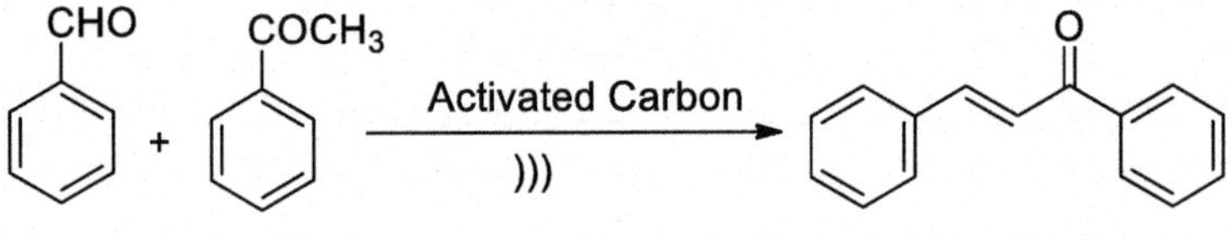

Scheme 2.18 Synthesis of chalcone promoted by sonication and activated carbon

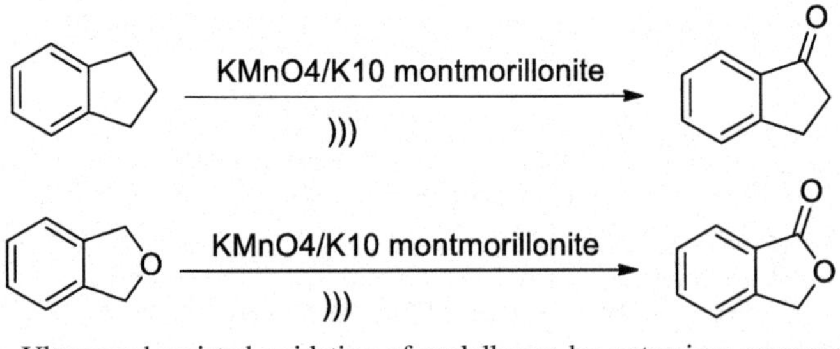

Scheme 2.19 Ultrasound-assisted oxidation of arylalkenes by potassium permanganate on K10 montmorillonite

2.2 Ultrasound in Synthetic Organic Chemistry

derivatives under low-frequency ultrasound (Scheme 2.20). Sonication assists in reducing the particle size and improving the specific surface area of powdered zinc, which is ultimately responsible for enhancing the catalytic reduction of the substrate (Peng et al. 2005).

In another study, Wang et al. reported the chemoselective reduction of the olefinic double bond of coumarin without opening the ring was achieved with the help of ultrasonic irradiations. Catalytic hydrogenation of coumarin with the Raney nickel and H_2 gas can produce the hydrogenated coumarin. Similar effects were also observed in the α, β-unsaturated ketones, where only double bonds were reduced without touching ketones (Scheme 2.21). The mechanical effects of sonochemistry on the Raney nickel dominate during the reaction (Wang et al. 1999).

Aromatic nitro compounds have also been reduced using Pd/C and hydrogen sources into respective aromatic amines in water and a 2-MeTHF mixture under ultrasonic irradiation (Scheme 2.22). The convention synthesis achieved this conversion in 90 min, while the same was carried out in 15 min through the sonication. The effective mass transfer was achieved by physical effect through the sonication responsible for speeding up the reaction (Letort et al. 2017).

The sonication effect of the ultrasonic horn has also sped up the Barbier reaction. In this instance, the benzaldehyde and vinyl/alkynyl halide model substrate reacted conventionally in the presence of zinc THF/NH_4Cl solvent, but it took a longer reaction time

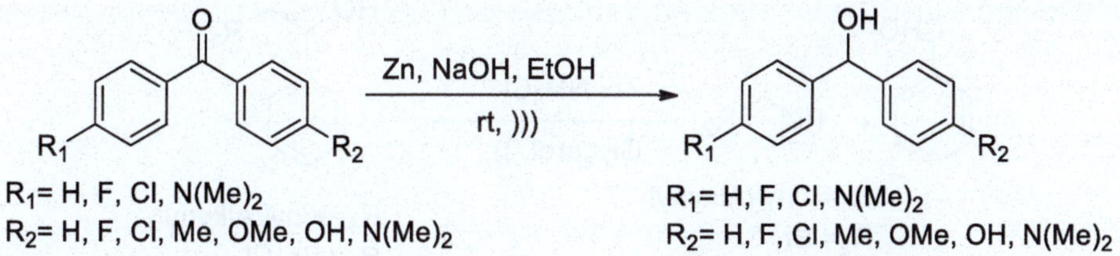

R_1= H, F, Cl, N(Me)$_2$
R_2= H, F, Cl, Me, OMe, OH, N(Me)$_2$

R_1= H, F, Cl, N(Me)$_2$
R_2= H, F, Cl, Me, OMe, OH, N(Me)$_2$

Scheme 2.20 Reduction of benzophenone with zinc under ultrasound irradiation

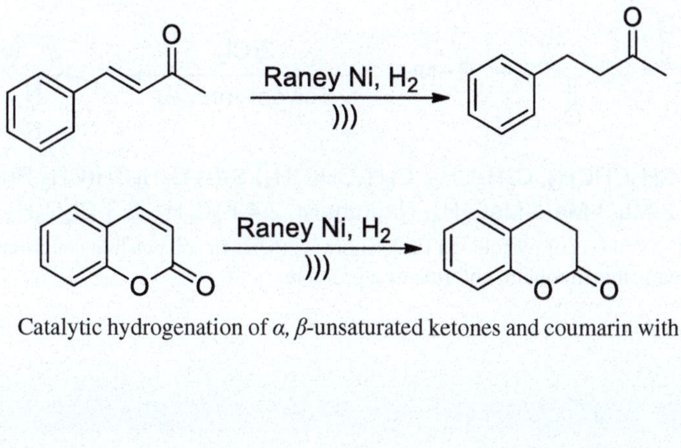

Scheme 2.21 Catalytic hydrogenation of α, β-unsaturated ketones and coumarin with Raney nickel catalyst

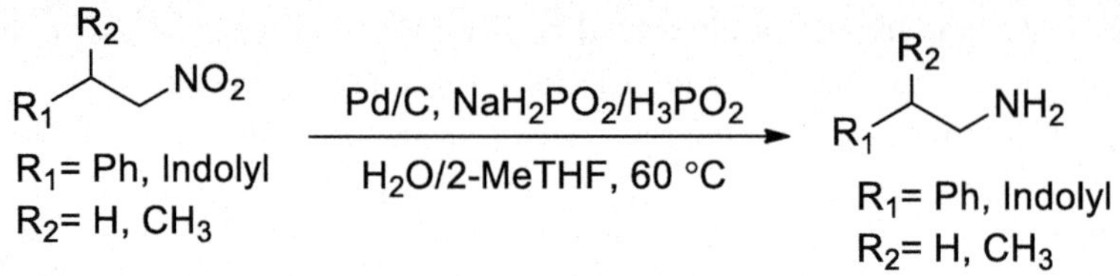

Scheme 2.22 Reduction of nitroaromatic via sonication

(90 min) and lower yield. The same reaction was completed in 15 min using an ultrasonic horn without any byproduct formation (Scheme 2.23) (Cravotto et al. 2011).

Zhang et al. synthesized the pyrrole by cyclizing diketones and aromatic amines via Paal-Knorr synthesis in solvent-free conditions under ultrasonic irradiation (Scheme 2.24). Zirconium chloride was used as a promotor for the reaction (Zhang et al. 2008).

The carbon–carbon coupling Heck reaction of various electron-deficient terminal alkene with 3-iodo-1-methyl-1H-indole under ultrasound irradiation catalyzed by palladium on carbon, triphenylphosphine, and triethyl amine (Scheme 2.25). The 3-vinylindole obtained a good to excellent yield (Bhavani et al. 2019).

The long-chain compounds have also been synthesized with the help of sonochemistry. In this regard, Pandit et al. synthesized various substituted benzoxazoles from o-aminocardanol and aldehydes using Indion 190 resin as a heterogeneous catalyst

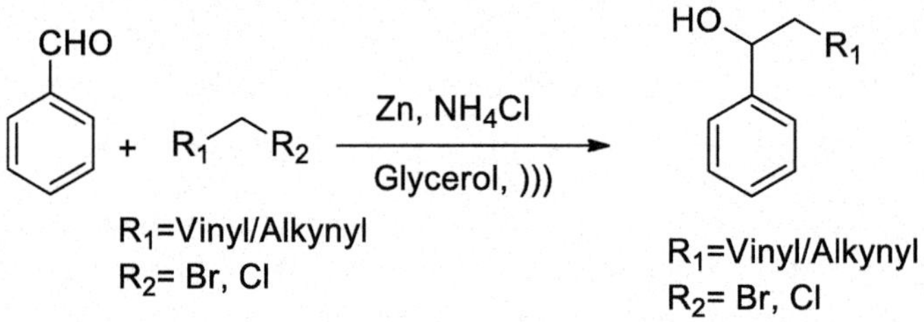

Scheme 2.23 Ultrasound-assisted Barbier reaction

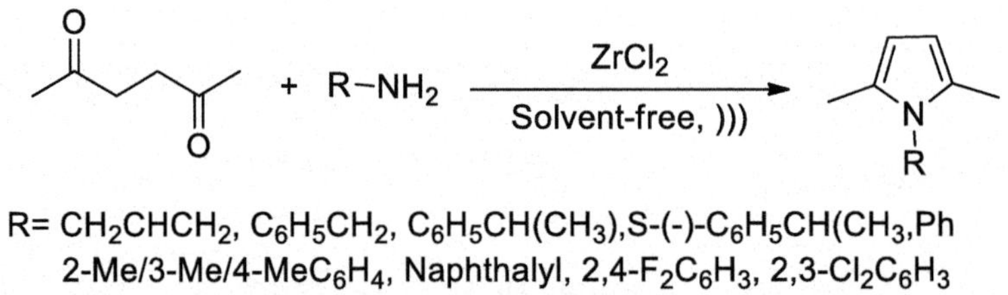

R= CH_2CHCH_2, $C_6H_5CH_2$, $C_6H_5CH(CH_3)$, S-(-)-$C_6H_5CH(CH_3)$, Ph
2-Me/3-Me/4-MeC_6H_4, Naphthalyl, 2,4-$F_2C_6H_3$, 2,3-$Cl_2C_6H_3$

Scheme 2.24 Synthesis of pyrrole via Paal-Knorr synthesis in solvent-free conditions under ultrasonic irradiation using zirconium chloride as a promoter

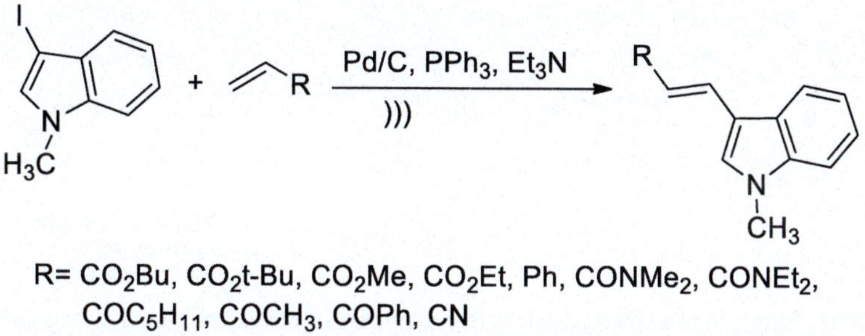

2.2 Ultrasound in Synthetic Organic Chemistry

R= CO_2Bu, $CO_2t\text{-}Bu$, CO_2Me, CO_2Et, Ph, $CONMe_2$, $CONEt_2$, COC_5H_{11}, $COCH_3$, COPh, CN

Scheme 2.25 Ultrasound-assisted Heck reaction using Pd/C, triphenylphosphine, and triethyl amine

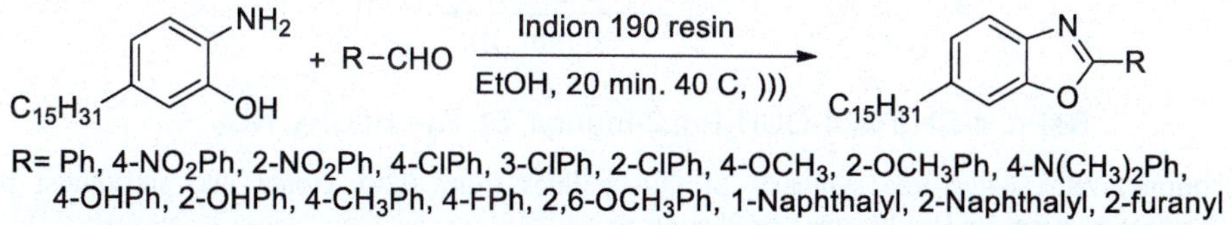

R= Ph, 4-NO_2Ph, 2-NO_2Ph, 4-ClPh, 3-ClPh, 2-ClPh, 4-OCH_3, 2-OCH_3Ph, 4-$N(CH_3)_2$Ph, 4-OHPh, 2-OHPh, 4-CH_3Ph, 4-FPh, 2,6-OCH_3Ph, 1-Naphthalyl, 2-Naphthalyl, 2-furanyl

Scheme 2.26 Heterogenous synthesis of substituted benzoxazoles by Indion 190 resin as a heterogeneous catalyst under ultrasonic irradiations

(Scheme 2.26). Multiple novel compounds were prepared under mild, green, and eco-friendly protocols, which can be further explored for antitubercular activity (Patil et al. 2023).

2.2.3 Sonocatalytic Reaction

The sonocatalytic reactions are the reactions that are catalyzed only in the presence of ultrasound waves. The output from the sonochemical effects and the conventional heating differ in multiple reactions. Hence, discussing such methods in a separate section with probable ultrasonic effects is paramount. The following protocols only utilize ultrasonic irradiation to prepare the various synthetic compounds.

Catalyst-free synthesis of novel derivatives of the isatin-hydrazinothiazole from substituted isatin, aryl α-bromoketones, and thiosemicarbazide with the help of ultrasound (Scheme 2.27). Furthermore, those novel compounds are evaluated for their antioxidant activity (Khanum and Pasha 2024).

Multiple quinoxaline derivatives were synthesized from 1,2-diketones and substituted o-phenylenediamines with catalyst-free conditions under ultrasonic irradiation (Scheme 2.28). The present reaction condition tolerated various aromatic and heteroaromatic diketones. The negligible effect of electron-donating and withdrawing attacking amines was observed on the yield (Guo et al. 2009).

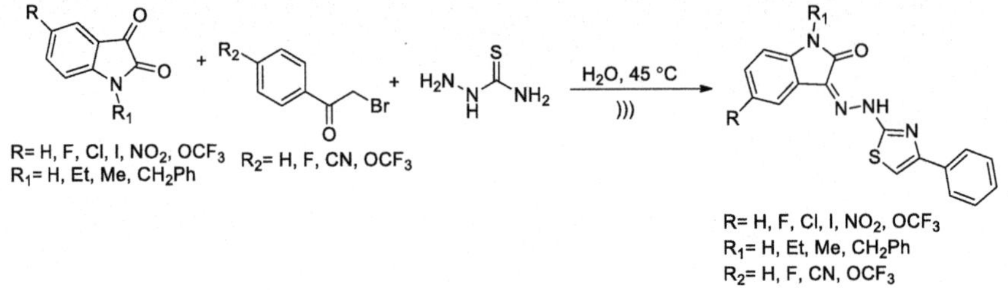

Scheme 2.27 Ultrasound-assisted catalyst-free synthesis of novel isatin-hydrazinothiazoles

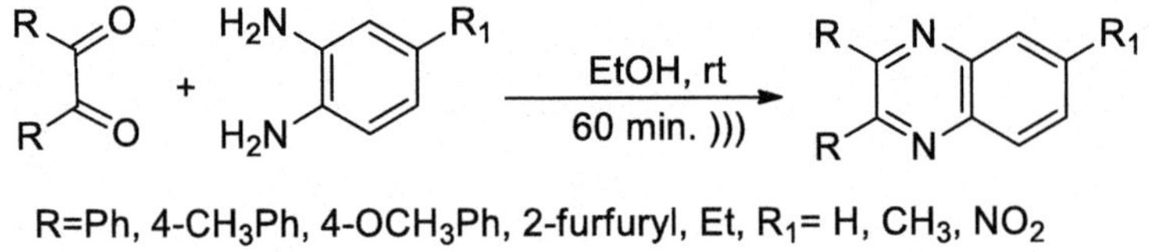

R=Ph, 4-CH$_3$Ph, 4-OCH$_3$Ph, 2-furfuryl, Et, R$_1$= H, CH$_3$, NO$_2$

Scheme 2.28 Catalyst-free synthesis of quinoxalines from 1,2-diketones and substituted o-phenylenediamines under ultrasonic irradiation

Sonochemical synthesis of α-sulfamidophosphonates was achieved under catalyst-free conditions (Scheme 2.29). This tree-component reaction was carried out from the starting material of aryl/alkyl sulfonamide, triethyl phosphoramide, and substituted aldehydes. The mild reaction condition, good to excellent yields, eco-friendly synthesis, and easy isolation procedure are some of the notable advantages of the current method (Belhani et al. 2015).

Aouf et al. successfully protected the different alkyl, aryl, and heterocyclic amines by using *di-tert*-butyl dicarbonate ((Boc)$_2$O) with the aid of ultrasound (Scheme 2.30). This catalyst and solvent-free chemoselective approach has many advances, including cost-effective and easily available starting materials, shorter reaction times, and milder reaction conditions (Amira et al. 2014).

Cunico et al. have utilized ultrasound for efficient multicomponent synthesis to afford 2-aryl-3-(piperonylmethyl)-1,3-thiazolidin-4-ones from piperonilamine, mercaptoacetic acid, and arylaldehydes. This protocol offers a shorter reaction time, broad applicability

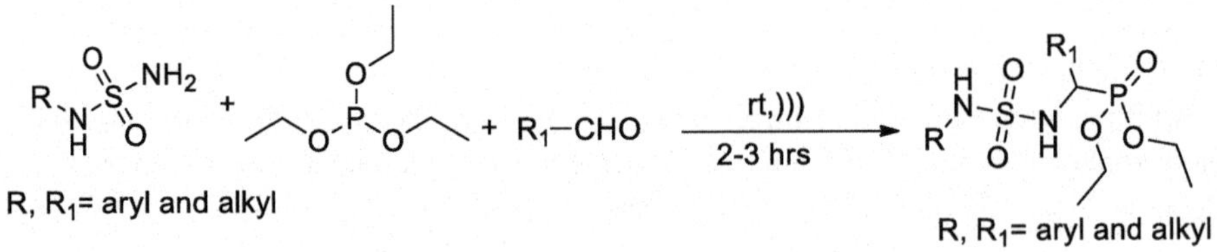

Scheme 2.29 Sonochemical synthesis of α-sulfamidophosphonates under catalyst-free condition

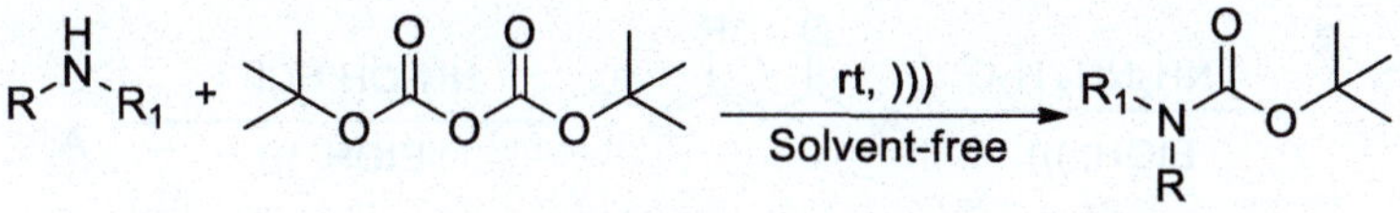

2.2 Ultrasound in Synthetic Organic Chemistry

R and R_1= H, Ph, 4-OCH_3Ph, 2-OMePh, 4-OHPh, $PhCH_2$, $PhCHCH_2$, cyclohexyl, propyl, diPh, piprazinaylphenyl, tetrahydroquinolyl,

Scheme 2.30 Ultrasound-assisted protection of various amines with (($Boc)_2O$)

to aryl and heteroaryl aldehydes, and a good yield of the desired 1,3-thiazolidin-4-ones (Neuenfeldt et al. 2011) (Scheme 2.31).

Jonnalagadda et al. developed the catalyst-free protocol for efficient, mild, and green synthesis of dihydroquinolines from the substituted aldehydes, malononitrile and resorcinol/naphthol in water under ultrasonic irradiation (Scheme 2.32). The present method has various advantages, such as catalyst-free protocol, high yield of desired products, simple workup procedure, use of aqueous solvent, and readily available starting materials (Pagadala et al. 2014).

Pyrazole and isoxazole are two important heterocycles that were synthesized using enaminones with hydrazine hydrate and hydroxylamine hydrochloride in ethanol by ultrasound-mediated synthesis without any catalyst (Scheme 2.33). This eco-friendly and high-yielding protocol also offers simple isolation of desired products and a comparatively shorter reaction time (Huang et al. 2014).

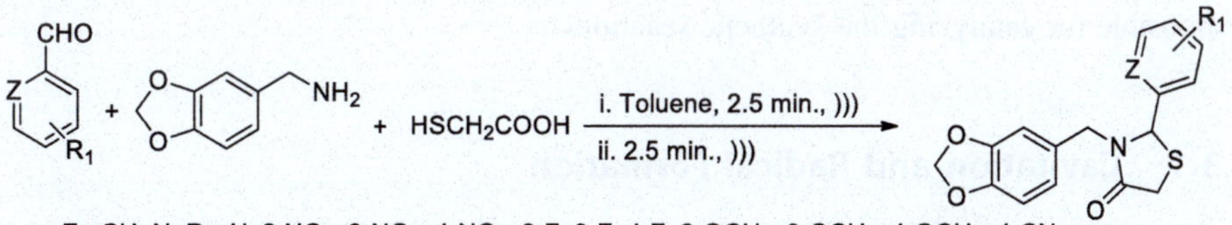

Z= CH, N; R_1=H, 2-NO_2, 3-NO_2, 4-NO_2, 2-F, 3-F, 4-F, 2-OCH_3, 3-OCH_3, 4-OCH_3, 4-CN

Scheme 2.31 Catalyst-free multicomponent synthesis of 2-aryl-3-(piperonylmethyl)-1,3-thiazolidin-4-ones from piperonilamine, mercaptoacetic acid, and arylaldehydes under ultrasonic irradiations

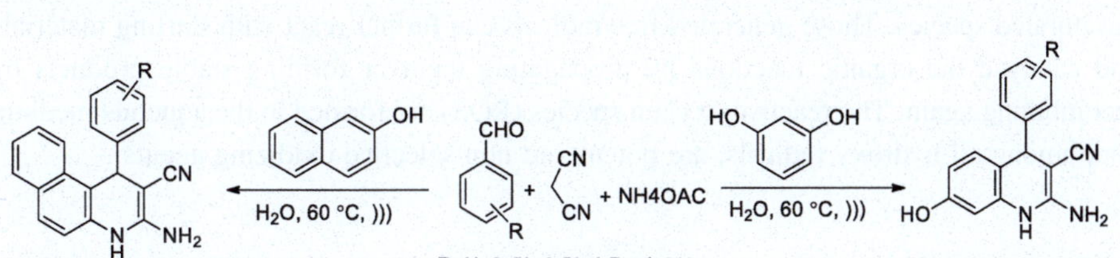

R=H, 2-Cl, 4-Cl, 4-OH

R=H, 2-Cl, 4-Cl, 4-Br, 4-OH

R=H, 4-Cl, 4-Br, 4-OH

Scheme 2.32 Catalyst-free synthesis of dihydroquinolines from the substituted aldehydes, malononitrile and resorcinol/naphthol in water under ultrasonic irradiation

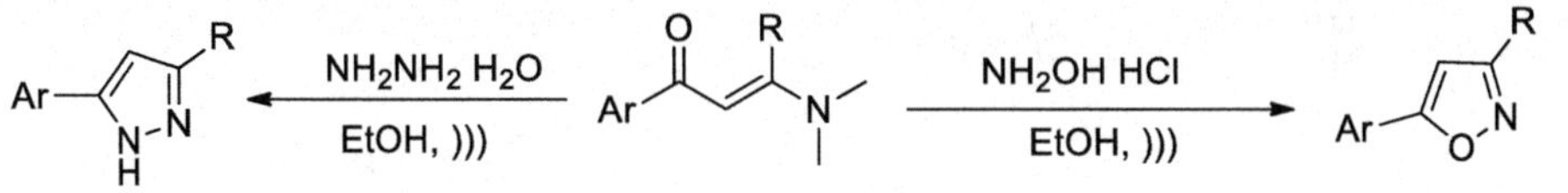

Ar= Ph, 4-ClPh,4-BrPh, 2,4-diClPh, 4-CH_3OPh, 4-BocNHPh, Naphthalene-1-yl, 4-CH_3Ph
R= H, CH_3

Scheme 2.33 Synthesis of pyrazoles and isoxazoles in catalyst-free conditions using enaminones with hydrazine hydrate and hydroxylamine hydrochloride in ethanol by ultrasound-mediated synthesis

2.3 Organic Sonochemistry: Mechanism and Reactivity

As discussed earlier, the sonochemical effect of ultrasounds promoted many reactions, including cyclization, hydrolysis, substitution, oxidation–reduction, and carbon–carbon coupling reactions. Understanding the selectivity resulting from sonochemistry requires knowledge of the mechanism behind organic sonochemistry. Jean-Louis divided the sonochemical reaction into three types in 1990: (a) Type I reactions occurred in the homogenous medium, and the free radicals generations via cavitation are primarily responsible for the reaction. This is also termed "true sonochemistry"; (b) Type II reactions mainly take place in the heterogeneous interfaces through a mechanical effect from the sonic waves; This is also termed "false sonochemistry"; (c) Type III reactions depend solely on single electron transfer, a critical step (Einhorn et al. 1990; Luche et al. 1990). Apart from this, the following effects from the ultrasounds are also reported and responsible for catalyzing the synthetic reactions.

2.3.1 Cavitation and Radical Formation

As discussed in Chap. 1, ultrasound profoundly impacts the liquid medium to develop bubbles or microsized cavities. When those bubbles grow and collapse, the quasi-adiabatic process increases the temperature by nearly 1000 K and pressure to a thousand bars inside the bubble. Cavitation in homogenous media generates free radicals or coordinatively unsaturated species. Those generated free radicals can further react with starting materials and catalyze the organic reactions by propagating them or forming stable products by recombining again. The reactive oxygen species (ROS) are formed in the aqueous medium and, among all hydroxy radicals, are potent and non-selective oxidizing agents.

2.3.2 Mechanical Effects

Due to the ultrasound, the bubbles grow and collapse, creating shock waves and shear forces that can break non-volatile molecules accumulated at the bubble interface. Mass and energy transfer improvements have resulted from the high-velocity flows and quartz wind at the solid–liquid and liquid–liquid interfaces in the heterogeneous reaction. In many cases, the reaction process is through the ionic intermediates generated from the mechanical effects of ultrasonic irradiations. Also, the implosion of bubbles on surfaces generates microjets that lead to powerful hammering and interparticle collisions. These processes significantly enhance the mechanical effect of acoustic cavitation and cleaning efficacy, demonstrating its valuable role in the degradation of various organic hydrophobic pollutants.

2.3.3 Reactivity and Sonochemical Switching

In 1996, Luche and Mason wrote a book on "Ultrasound as a New Tool for Synthetic chemists" in a book "Chemistry Under Extreme or non-Classical Conditions" edited by van R. V. Eldik R., Hubbard C. D. (Mason and Luche 1996) In this book, the authors explained three governing rules for the reactivity of the reaction proceeded with the aid of sonochemistry, which are as follows:

Rule 1 applies to the reactions sensitive to ultrasonic effects in a homogenous medium and driven only through the radical mechanism. The reactions driven through ionic intermediates do not have any direct impact by ultrasonic irradiations.

Rule 2 is applicable to the heterogeneous medium where the reaction proceeds through the ionic intermediates generated by the mechanical effect of cavitation. Similar results would be expected if ion generation is achieved via mechanical agitation but may vary on a case basis.

Rule 3 applies to heterogeneous systems with a combined mechanism, i.e., the reactions involved in forming radicals and ions. Ultrasonic irradiation will enhance the radical component, yet a mechanical effect may still apply. Possible two situations were observed in those cases: (i) The first one is the "convergent process," where the formation of the same product and an increase in the overall rate of reaction result from both mechanisms; (ii) the different product may be obtained from the radical and ionic pathways but the **sonochemical switching** observed due to the sonication and the primary product only forms from the radical pathway. This is called "divergent processes" where ultrasound waves form the product chemoselectively via a radical mechanism over the ionic. So far, no new sonochemistry rules have been established besides those established by Jean-Louis Luche.

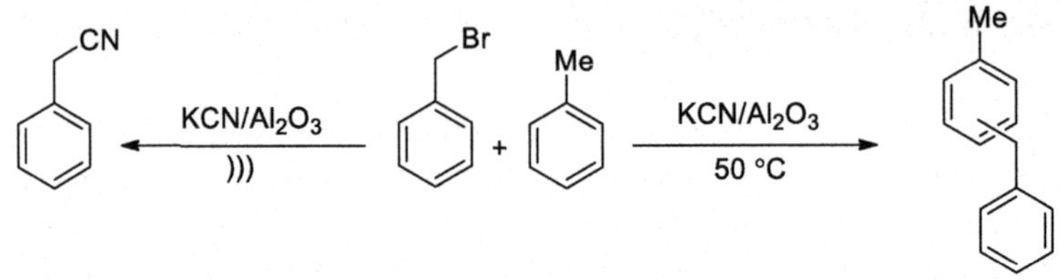

Scheme 2.34 Sonochemical switch reaction of benzyl bromide and toluene

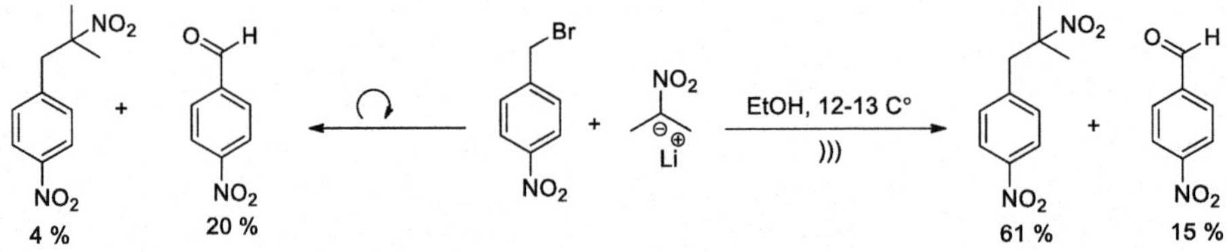

Scheme 2.35 Kornblum-Russell alkylation in polar and sonochemical conditions

The divergent process was first observed by Ando et al. in 1984; they synthesized the benzyl cyanide from the benzyl bromide in the presence of Al_2O_3-supported KCN as a catalyst in toluene suspension under ultrasonic irradiation (45 kHz) while the conventional heating at the 50 °C produces the diphenylmethane through the Friedel–crafts reaction (Scheme 2.34). The reaction mechanism was changed by sonication from the aromatic electrophilic substitution to the aliphatic nucleophilic substitution (Ando et al. 1984).

Another example of sonochemical switching is the Kornblum-Russell reaction between the 4-nitrobenzyl bromide and lithium 2-nitropropan-2-ide, which gives the 4-nitrobenzaldehyde (20%) as a primary product via ionic mechanism. Under the sonication, the major compound is dinitro, i.e., 1-(2-methyl-2-nitropropyl)-4-nitrobenzene (61%) (Scheme 2.35). The C/O alkylation ratio is almost reversed in optimal irradiation conditions compared to the reaction without sonic waves, suggesting that ultrasounds have a direct intervention in the electron transfer process.

2.4 Application in Preparation of Inorganic Nanomaterials

Nanomaterials contain organic and inorganic structures with an ideal size below 100 nm (nm) or less (Ali Dheyab et al. 2021). These materials are made up of various elements like carbon, iron, titanium, silver, gold, silica, Zn, etc. They possess high surface areas, offering superior physiochemical properties over the micromaterials (Khan et al. 2019). Preparing those nanomaterials with the required morphology, size, specific surface area (SSA), and pore diameter properties has sparked significant interest. Various approaches have been exploited to prepare it, from simple-solution-based hydrothermal, solvothermal,

2.4 Application in Preparation of Inorganic Nanomaterials

and sonochemical methods to the most advanced lithography, aerosol, electrospinning, laser ablation, and epitaxy (Baig et al. 2021). Among all of the approaches, preparation via sonochemistry is very economical, and it provides unique control on crystallinity, allowing the synthesis of amorphous metal oxide and related materials. Therefore, the following are the nanomaterials prepared using sonochemical synthesis.

2.4.1 Metal Oxide Nanoparticles

Metal oxide nanoparticles have attracted the attention of chemists in the last two decades and have found applications in various fields, not limited to energy, electronics, pharmaceuticals, and biomedicals (Chavali and Nikolova 2019). Many of these materials exhibit lower porosity, with specific surface area (SSA) limits set at 100 m^2/g. Some studies report that metal oxides synthesized using a sonochemical synthetic strategy have relatively high porosity and improved SSA. Several metal oxide nanoparticles, including NiO, TiO_2, CeO_2, $CeVO_4$, SnO_2, $MgAl_2O_4$, CuO, ZnO, MnO_2, RuO_2, Fe_2O_3, and Al_2O_3, have been reported by utilizing ultrasound-assisted synthesis (Głowniak et al. 2023).

Recently, Pandele et al. synthesized the RuO_2 nanoparticles using ultrasonic horn. Further, these nanoparticles were decorated with multiwalled carbon nanotubes (MWCNTs) and confirmed by X-ray diffraction, Raman spectroscopy, and energy-dispersive X-ray analysis. Further, polyetherimine, as the polymer matrix, was used along with decorated RuO_2 nanoparticles to develop the membrane composite (Voicu et al. 2024).

CuO and Cu_2O nanoparticles were also synthesized under ultrasound irradiations and evaluated against the copper-resistant pseudomonas strains (Havryliuk et al. 2024).

In another study, sonochemical synthesis of ZnO/Bi_2O_3 metal oxide nanophotocatalyst was achieved and utilized for diesel fuel's photocatalytic desulfurization (Mahmoud et al. 2023).

2.4.2 Silicas and Organosilica

Silicas and organosilica have been exploited in many research areas, including electronics and photonics, catalysis, drug delivery, biosensing, and wastewater treatment for the removal of pollutants and heavy metals (Yang et al. 2019; Osta et al. 2020). The hydrolysis of silica sources generally results in the forming a polymeric network. The incorporation of organic compounds into the silica network may change the product's structure or result in an undesired product. Hence, sonochemical synthesis was used to introduce the additive in the silica to form ordered mesoporous silicas (Vetrivel et al. 2010; Osta et al. 2020). Palani et al. employed sonochemical synthesis to prepare mesoporous silica SBA-15 using a rapid synthesis method that requires less time. The prepared silica has a good

SSA and pore diameter (Palani et al. 2010). Kim et al. also prepared the silica nanoparticles using the sol–gel method under ultrasound irradiation. The synthesis was done in the solvent system (EtOH-H_2O) using tetraethyl orthosilicate as a precursor in the presence of ammonia at room temperature. The particle size ranged from 40 to 400 nm and had a spherical morphology characterized by scanning electron microscopy (SEM) (Kim et al. 2016).

2.4.3 Metal–Organic Frameworks (MOFs) and Covalent-Organic Frameworks (COFs)

Metal–organic frameworks (MOFs) are innovative porous materials developed from organic ligands and metal ions, forming versatile one-, two-, or three-dimensional structures. These unique properties make MOFs essential in various applications, including water purification, drug delivery applications, gas storage, and catalysis (Raptopoulou 2021). Recently, Khunt et al. synthesized glutamic acid-based MOFs through ultrasonic irradiation, and further, it is used for the adsorptive removal of a xylenol orange dye. This study prepared a series of MOFs using varying amounts of copper, zinc, and manganese acetate salt. The adsorption data fitted well in Langmuir isotherm, and adsorption follows pseudo-first-order kinetics (Bhanderi et al. 2025). An innovative method for the preparation of MOF was developed by Park et al. using zinc as a metal. This ultrasound-assisted solvothermal method was able to synthesize the MOFs in 2–180 min by consuming less than 0.03 mol of solvent at room temperature. The prepared MOFs have high crystallinity and porosity, indicating they can have diverse applications in material science, biomedical, and wastewater treatment (Yi et al. 2024).

Covalent-organic frameworks (COFs) are novel crystalline porous materials synthesized by linking organic units into structured 2D or 3D networks via covalent bonds. In this context, Wei et al. synthesized the novel COF 1,3,6,8-tetrakis(4-formylphenyl)pyrene-adipic dihydrazide (TFPPy-AD) from 1,3,6,8-tetrakis(4-formylphenyl)pyrene (TFPPy) and adipic dihydrazide (AP) monomers under ultrasonic irradiation at 60 °C in 1 h. The prepared COF was used for the selective adsorption of flavonoids with high adsorptive capacity and extremely rapid adsorption. This sonochemical method offers several advantages compared to the solvothermal method, such as a shorter duration, the use of less toxic materials, and the requirement of lower temperatures for synthesis (Wei et al. 2024). Thakore et al. prepared the ultrasound-assisted COF ellagic acid and diisocyanates through the urethane linkage. This bioinspired COF offers several advantages over traditional solvothermal methods, including one-pot synthesis, eco-friendly production, the use of biodegradable molecules as starting materials, and the incorporation of auxiliary solvents (Thakkar et al. 2025).

2.4.4 Carbons

Carbon-based materials (CBMs) include biochar, activated carbon, carbon nanotubes, fullerenes, carbon dots, nanodiamonds, and graphitic carbonitride. They have potential applications in water treatment, CO_2 capture, drug delivery, catalysis, and sensing (Egbedina et al. 2022). Multiple methods were reported for the synthesis of CBMs, such as pyrolysis, hydrothermal method, laser ablation, precipitation, combustion, and the sonochemical method (Rao et al. 2021). The drawbacks of the above methods are cytotoxicity from chemical methods due to the use of non-safe chemicals, precipitation having low-yielding protocols, the requirement of high temperature in pyrolysis, and the high cost associated with laser ablation. The sonochemical method is simple, convenient, economical, and green compared to all the above methods (Mason 2007).

In this context, Gedanken et al. prepared the gallium-doped carbon dots with the help of ultrasonic irradiations and then evaluated them for antibacterial activity against the *pseudomonas aeruginosa* (Mason 2007). In addition, Kang et al. synthesized the carbon quantum dots (CQDs) through sonication and such CQDs were able to catalyze the aldol reaction by using aromatic aldehydes and acetone. The excellent yield of the products (89%) was obtained from this nanocatalyst when the 4-formylbenzonitrile was used as a starting material (Han et al. 2014). Figure 2.1 summarizes the applications of sonochemistry in the preparation of nanomaterials.

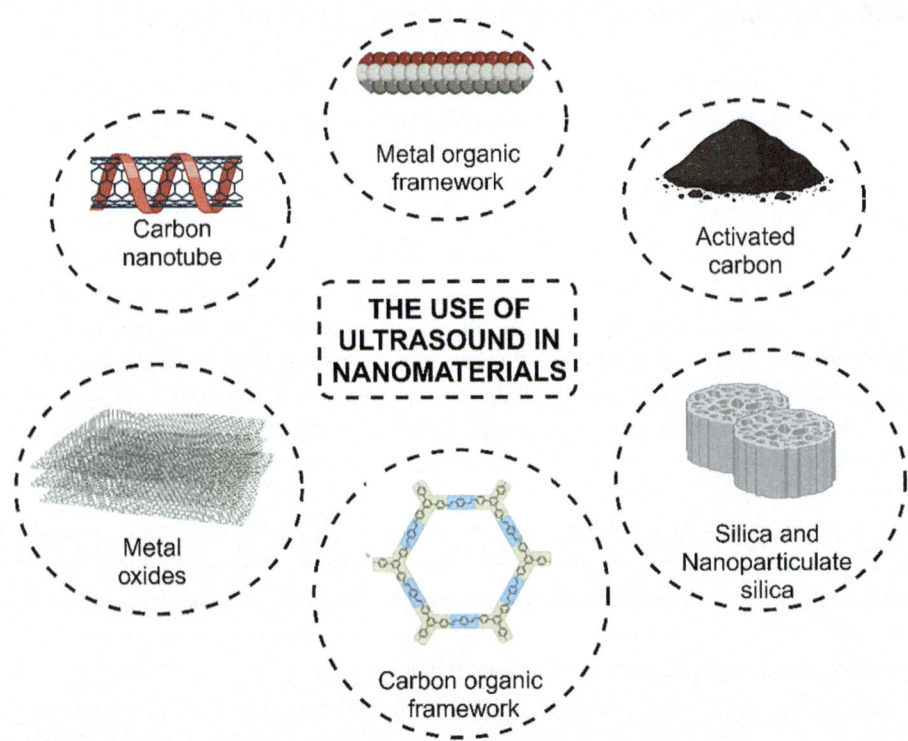

Fig. 2.1 Preparation of nanomaterials using sonochemistry (created by biorender.com)

2.5 Conclusion

The ultrasonic-assisted synthesis of homogeneous, heterogeneous organic products as well as inorganic nanomaterials has garnered significant interest from chemists due to this method being mild, convenient, eco-friendly, green, and economical. The ultrasonic waves have catalyzed many reactions by achieving high temperature and pressure within nanoseconds in the microbubbles. In addition, ultrasound can be able to initiate free radical generation via the lysis of the medium. The secondary mechanical effects, such as shock waves or microjet formation, catalyze the heterogeneous reactions. However, its applications are not limited to organic synthesis; varieties of inorganic nanomaterials, such as metal oxides, mesoporous silica, MOF, COF, and carbon-based materials, were prepared through the sonochemical synthesis. In summary, sonochemical synthesis is a widely utilized approach to chemical synthesis, and substantial work is needed to design new reactors to achieve the scale for sustainability.

References

Adapa S et al (2023) Ultrasound assisted synthesis of 4-(1H-indol-3-yl)thieno[2,3-d]pyrimidine derivatives via AcOH mediated C C bond forming reaction. Tetrahedron Lett 131:154784. https://doi.org/10.1016/j.tetlet.2023.154784

Ahmad MJ et al (2016) Synthesis, in vitro potential and computational studies on 2-amino-1,4-dihydropyrimidines as multitarget antibacterial ligands. Med Chem Res 25(9):1877–1894. https://doi.org/10.1007/s00044-016-1613-z

Ali Dheyab M, Aziz AA, Jameel MS (2021) Recent advances in inorganic nanomaterials synthesis using sonochemistry: a comprehensive review on iron oxide, gold and iron oxide coated gold nanoparticles. Molecules 26(9):2453. https://doi.org/10.3390/molecules26092453

Amira A et al (2014) A simple, rapid, and efficient N-Boc protection of amines under ultrasound irradiation and catalyst-free conditions. Monatshefte Für Chemie—Chemical Monthly 145(3):509–515. https://doi.org/10.1007/s00706-013-1094-4

Ando T et al (1984) Sonochemical switching of reaction pathways in slid–liquid two-phase reactions. J Chem Soc Chem Commun 7:439–440. https://doi.org/10.1039/C39840000439

Baig N, Kammakakam I, Falath W (2021) Nanomaterials: a review of synthesis methods, properties, recent progress, and challenges. Mater Adv 2(6):1821–1871. https://doi.org/10.1039/D0MA00807A

Bao J et al (2023) Sonoactivated nanomaterials: a potent armament for wastewater treatment. Ultrason Sonochem 99:106569. https://doi.org/10.1016/j.ultsonch.2023.106569

Belhani B et al (2015) A one-pot three-component synthesis of novel α-sulfamidophosphonates under ultrasound irradiation and catalyst-free conditions. RSC Adv 5(49):39324–39329. https://doi.org/10.1039/C5RA03473F

Bhanderi PM et al (2025) Facile sonochemical synthesis and characterization of amino acid-based metal-organic frameworks for enhanced dye adsorption. J Mol Struct 1326:141058. https://doi.org/10.1016/j.molstruc.2024.141058

Bhavani S et al (2019) Ultrasound assisted Mizoroki-Heck reaction catalyzed by Pd/C: Synthesis of 3-vinyl indoles as potential cytotoxic agents. Arab J Chem 12(8):3836–3846. https://doi.org/10.1016/j.arabjc.2016.02.002

Bhowmik D et al (2024) An efficient, green and micellar catalyzed preparative-scale synthesis of 3,4-Disubstituted isoxazole-5(4H)-ones in the water. Sustain Chem Environ 5:100070. https://doi.org/10.1016/j.scenv.2024.100070

Calvino V et al (2006) Ultrasound accelerated Claisen-Schmidt condensation: a green route to chalcones. Appl Surf Sci 252(17):6071–6074. https://doi.org/10.1016/j.apsusc.2005.11.006

Chatel G, Colmenares JC (2017) Sonochemistry: from basic principles to innovative applications. Top Curr Chem 375(1):8. https://doi.org/10.1007/s41061-016-0096-1

Chavali MS, Nikolova MP (2019) Metal oxide nanoparticles and their applications in nanotechnology. SN Appl Sci 1(6):607. https://doi.org/10.1007/s42452-019-0592-3

Cintas P, Luche J-L (1999) Green chemistry. Green Chem 1(3):115–125. https://doi.org/10.1039/a900593e

Cravotto G et al (2011) Efficient synthetic protocols in glycerol under heterogeneous catalysis. Chemsuschem 4(8):1130–1134. https://doi.org/10.1002/cssc.201100106

Dandia A, Gupta SL, Parewa V (2014) An efficient ultrasound-assisted one-pot chemoselective synthesis of pyrazolo[3,4-b] pyridine-5-carbonitriles in aqueous medium using NaCl as a catalyst. RSC Adv 4(14):6908. https://doi.org/10.1039/c3ra47231k

Daroughehzadeh Z, Kiyani H (2024) Arylideneisoxazole-5(4 H)-one synthesis by organocatalytic three-component hetero-cyclization. Polycyclic Aromat Compd 44(5):3200–3221. https://doi.org/10.1080/10406638.2023.2231602

Draye M, Estager J, Kardos N (2019) Organic sonochemistry: ultrasound in green organic synthesis. In: Activation methods. Wiley, pp 1–93. https://doi.org/10.1002/9781119687443.ch1

Egbedina AO et al (2022) Emerging trends in the application of carbon-based materials: a review. J Environ Chem Eng 10(2):107260. https://doi.org/10.1016/j.jece.2022.107260

Einhorn C et al (1990) Organic sonochemistry—some illustrative examples of a new fundamental approach. Tetrahedron Lett 31(29):4129–4130. https://doi.org/10.1016/S0040-4039(00)97560-9

Fujita M et al (2015) Sono-bromination of aromatic compounds based on the ultrasonic advanced oxidation processes. Ultrason Sonochem 27:247–251. https://doi.org/10.1016/j.ultsonch.2015.04.030

Główniak S et al (2023) Recent developments in sonochemical synthesis of nanoporous materials. Molecules 28(6):2639. https://doi.org/10.3390/molecules28062639

Guo W-X et al (2009) An efficient catalyst-free protocol for the synthesis of quinoxaline derivatives under ultrasound irradiation. J Braz Chem Soc 20(9):1674–1679. https://doi.org/10.1590/S0103-50532009000900016

Han Y et al (2014) Carbon quantum dots with photoenhanced hydrogen-bond catalytic activity in aldol condensations. ACS Catal 4(3):781–787. https://doi.org/10.1021/cs401118x

Handore KN et al (2021) Ultrasound-assisted solvent-free synthesis of 3, 4-dihydropyrimidin-2(1 H)-ones/thiones using polyindole as a recyclable catalyst. Polymer-Plastics Technol Mater 60(3):306–315. https://doi.org/10.1080/25740881.2020.1811313

Hatvate NT, Ghodse SM (2020) One-pot three-component synthesis of isoxazole using ZSM-5 as a heterogeneous catalyst. Synth Commun 50(23):3676–3683. https://doi.org/10.1080/00397911.2020.1815786

Havryliuk O et al (2024) Unveiling the potential of CuO and Cu_2O nanoparticles against novel copper-resistant pseudomonas strains: an in depth comparison. Nanomaterials 14(20):1644. https://doi.org/10.3390/nano14201644

Huang Z et al (2014) An efficient synthesis of isoxazoles and pyrazoles under ultrasound irradiation. J Heterocyclic Chem 51(S1). https://doi.org/10.1002/jhet.2016

Jin H et al (2009) Efficient epoxidation of chalcones with urea-hydrogen peroxide under ultrasound irradiation. Ultrason Sonochem 16(3):304–307. https://doi.org/10.1016/j.ultsonch.2008.10.013

Khan I, Saeed K, Khan I (2019) Nanoparticles: properties, applications and toxicities. Arab J Chem 12(7):908–931. https://doi.org/10.1016/j.arabjc.2017.05.011

Khanum A, Pasha MA (2024) Catalyst-free synthesis of new hydrazino thiazole derivatives in water under ultrasonication and evaluation of their antioxidant activity. Tetrahedron 167:134227. https://doi.org/10.1016/j.tet.2024.134227

Kim H-M, Lee C-H, Kim B (2016) Sonochemical synthesis of silica particles and their size control. Appl Surf Sci 380:305–308. https://doi.org/10.1016/j.apsusc.2015.12.048

Letort S et al (2017) New insights into the catalytic reduction of aliphatic nitro compounds with hypophosphites under ultrasonic irradiation. Green Chem 19(19):4583–4590. https://doi.org/10.1039/C7GC01622K

Li Z et al (2021) Sonochemical catalysis as a unique strategy for the fabrication of nano-/microstructured inorganics. Nanoscale Adv 3(1):41–72. https://doi.org/10.1039/D0NA00753F

Luche JL et al (1990) Organic sonochenistry : A new interpretation and its consequences. Tetrahedron Lett 31(29):4125–4128. https://doi.org/10.1016/S0040-4039(00)97559-2

Machado IV et al (2021) Greener organic synthetic methods: sonochemistry and heterogeneous catalysis promoted multicomponent reactions. Ultrason Sonochem 78:105704. https://doi.org/10.1016/j.ultsonch.2021.105704

References

Mahmoud RMA et al (2023) Sonochemical synthesis of heterostructured ZnO/Bi$_2$O$_3$ for photocatalytic desulfurization. Sci Rep 13(1):19391. https://doi.org/10.1038/s41598-023-46344-0

Martínez RF, Cravotto G, Cintas P (2021) Organic sonochemistry: a chemist's timely perspective on mechanisms and reactivity. J Org Chem 86(20):13833–13856. https://doi.org/10.1021/acs.joc.1c00805

Mason TJ (2007) Sonochemistry and the environment – Providing a "green" link between chemistry, physics and engineering. Ultrason Sonochem 14(4):476–483. https://doi.org/10.1016/j.ultsonch.2006.10.008

Mason TJ, Luche J-L (1996) Ultrasound as a new tool for synthetic chemists (edited by Rudi van Eldik)

Neuenfeldt PD et al (2011) Efficient sonochemical synthesis of thiazolidinones from piperonilamine. Ultrason Sonochem 18(1):65–67. https://doi.org/10.1016/j.ultsonch.2010.07.008

Osta O et al (2020) Direct synthesis of mesoporous organosilica and proof-of-concept applications in lysozyme adsorption and supported catalysis. ACS Omega 5(30):18842–18848. https://doi.org/10.1021/acsomega.0c01996

Pagadala R, Maddila S, Jonnalagadda SB (2014) Ultrasonic-mediated catalyst-free rapid protocol for the multicomponent synthesis of dihydroquinoline derivatives in aqueous media. Green Chem Lett Rev 7(2):131–136. https://doi.org/10.1080/17518253.2014.902505

Pagadala R, Kasi V, Perugu S (2020) Iodine-catalyzed ultrasound-assisted construction of pyridines and their glutamine synthetase molecular docking. J Chin Chem Soc 67(7):1296–1302. https://doi.org/10.1002/jccs.201900389

Palani A et al (2010) Rapid temperature-assisted sonochemical synthesis of mesoporous silica SBA-15. Microporous Mesoporous Mater 131(1–3):385–392. https://doi.org/10.1016/j.micromeso.2010.01.020

Patel PJ et al (2024) Sonochemical synthesis of benzylidene derivatives of enolizable carbonyls and their analogues in aqueous ethanol. Res Chem Intermed 50(3):1231–1248. https://doi.org/10.1007/s11164-023-05168-3

Patil BR et al (2023) Ultrasound-assisted facile and efficient synthesis of novel Benzoxazole derivatives from o-aminocardanol using Indion 190 resin as a reusable catalyst. J Chem Sci 135(1):4. https://doi.org/10.1007/s12039-022-02120-7

Peng Y, Zhong W, Song G (2005) Efficient and mild room temperature reduction of benzophenones under ultrasound irradiation. Ultrason Sonochem 12(3):169–172. https://doi.org/10.1016/j.ultsonch.2003.12.002

Pirsaheb M, Moradi N (2020) Sonochemical degradation of pesticides in aqueous solution: investigation on the influence of operating parameters and degradation pathway—a systematic review. RSC Adv 10(13):7396–7423. https://doi.org/10.1039/C9RA11025A

Pirsaheb M, Moradi N, Hossini H (2023) Sonochemical processes for antibiotics removal from water and wastewater: A systematic review. Chem Eng Res des 189:401–439. https://doi.org/10.1016/j.cherd.2022.11.019

Rao N, Singh R, Bashambu L (2021) Carbon-based nanomaterials: synthesis and prospective applications. Mater Today: Proc 44:608–614. https://doi.org/10.1016/j.matpr.2020.10.593

Raptopoulou CP (2021) Metal-organic frameworks: synthetic methods and potential applications. Materials 14(2):310. https://doi.org/10.3390/ma14020310

Rosales Pérez A, Esquivel Escalante K (2024) The evolution of sonochemistry: from the beginnings to novel applications. ChemPlusChem 89(6). https://doi.org/10.1002/cplu.202300660

Safaei-Ghomi J, Tavazo M, Mahdavinia GH (2018) Ultrasound promoted one-pot synthesis of 3,4-dihydropyrimidin-2(1H)-ones/thiones using dendrimer-attached phosphotungstic acid nanoparticles immobilized on nanosilica. Ultrason Sonochem 40:230–237. https://doi.org/10.1016/j.ultsonch.2017.07.015

Safari J, Gandomi-Ravandi S (2014) Application of the ultrasound in the mild synthesis of substituted 2,3-dihydroquinazolin-4(1H)-ones catalyzed by heterogeneous metal–MWCNTs nanocomposites. J Mol Struct 1072:173–178. https://doi.org/10.1016/j.molstruc.2014.05.002

Safari J, Gandomi-Ravandi S (2017) The combined role of heterogeneous catalysis and ultrasonic waves on the facile synthesis of 2,3-dihydroquinazolin-4(1H)-ones. J Saudi Chem Soc 21:S415–S424. https://doi.org/10.1016/j.jscs.2014.04.006

Shaabani A et al (2002) Selective oxidation of alkylarenes in dry media with potassium permanganate supported on Montmorillonite K10. Tetrahedron Lett 43(29):5165–5167. https://doi.org/10.1016/S0040-4039(02)00976-0

Shivanoori J et al (2024) Wang-OSO3H catalyzed a greener approach for the synthesis of imidazo[1,2-a]pyridin-3-amine derivatives as potential TNF-α inhibitors. J Mol Struct 1318:139280. https://doi.org/10.1016/j.molstruc.2024.139280

Song Y-L et al (2015) One-pot three-component synthesis of 3-hydroxy-5,5-dimethyl-2-[phenyl(phenylthio)methyl]cyclohex-2-enone derivatives under ultrasound. Ultrason Sonochem 22:119–124. https://doi.org/10.1016/j.ultsonch.2014.05.010

Thakkar H, Shukla F, Thakore S (2025) Ultrasound assisted synthesis of Ellagic acid derived bioinspired covalent organic framework via urethane polycondensation reaction. Appl Mater Today 42:102548. https://doi.org/10.1016/j.apmt.2024.102548

Valadi K et al (2020) Ultrasound-assisted synthesis of 1,4-dihydropyridine derivatives by an efficient volcanic-based hybrid nanocomposite. Solid State Sci 101:106141. https://doi.org/10.1016/j.solidstatesciences.2020.106141

Vetrivel S, Chen C-T, Kao H-M (2010) The ultrafast sonochemical synthesis of mesoporous silica MCM-41. New J Chem 34(10):2109. https://doi.org/10.1039/c0nj00379d

Voicu SI et al (2024) RUO 2 nanoparticle-decorated MWCNTS synthesized using a sonochemical method as reinforcing agents for PEI composite membranes. RSC Adv 14(53):39550–39558. https://doi.org/10.1039/D4RA07606K

Wang H et al (1999) Ultrasonic accelerated hydrogenation of α, β-unsaturated ketones with Raney nickel catalyst. Synth Commun 29(1):129–134. https://doi.org/10.1080/00397919908085744

Wei X et al (2024) Fast and facile sonochemical fabrication of covalent organic frameworks in water for the adsorption of flavonoids: adsorption behaviors and mechanisms. Colloids Surf, A 702:134731. https://doi.org/10.1016/j.colsurfa.2024.134731

Yang B, Chen Y, Shi J (2019) Mesoporous silica/organosilica nanoparticles: synthesis, biological effect and biomedical application. Mater Sci Eng R Rep 137:66–105. https://doi.org/10.1016/j.mser.2019.01.001

Yi J, Lee G, Park SS (2024) Solvent-induced structural rearrangement in ultrasound-assisted synthesis of metal–organic frameworks. Small Methods 8(12). https://doi.org/10.1002/smtd.202400363

Zhang Z-H, Li J-J, Li T-S (2008) Ultrasound-assisted synthesis of pyrroles catalyzed by zirconium chloride under solvent-free conditions. Ultrason Sonochem 15(5):673–676. https://doi.org/10.1016/j.ultsonch.2008.02.008

Sonochemistry in Chemical Engineering 3

Abstract

Sonochemistry, specifically dependent on acoustic cavitation, has transformed chemical engineering through its possibilities of provocation of localized temperature and pressure. This chapter will discuss sonochemical technology, specifically ultrasonic reactors, which include ultrasonic baths, horns, and multiple-frequency flow cell reactors. Bubble dynamics are modeled according to the Rayleigh-Plesset equation together with the role of frequency, temperature, pressure, liquid, reactor geometry, and the transducer's position. It has been used in nanoparticle preparation, nanoemulsion process, filtration process, ultrasonic atomization, and improvement of reaction rates in polymerization process, catalysis process, and enzymatic process. The chapter also discusses mass transfer modeling, namely the diffusion-limited model for vapor transport, ultrasound generation methods, piezoelectric and magnetostrictive transducers. Comparisons of sonochemistry with other hybrid technologies in controlling air contamination and biological wastewater treatment are also provided to highlight the importance of sonochemistry in developing more sustainable and innovative chemical engineering methods.

3.1 Introduction

Sonochemistry as a field of study combines ideas from physics, chemistry, and engineering as its areas of specialization. The physics aspects include the propagation through the fluids and the cavitation dynamics of the fluids, whereas the chemical impact involves the augmentation of the mechanisms and the reaction rate. In engineering terms, ultrasound benefits reactor design, process fine-tuning, and up-scaling from laboratory experiments (Asgharzadehahmadi et al. 2016). This coming together combines sonochemistry with the tenets of green chemistry, including waste avoidance, atoms economy, and energy utilization, making sonochemistry a key tool in sustainable engineering (Chatel 2018). Ultrasound epitomizes the balance of the four pillars by pursuing goals of combating resource scarcity and ecological decline in the global arena and by aligning science and technology with socio-economic goals.

Ultrasonic technologies could be applied to numerous chemical engineering processes and systems. In material synthesis and catalysis, ultrasound enables the enhancement of nanoparticles and sophisticated materials with accurate size and shape control (Sakthipandi et al. 2024). It is widely used in environmental engineering functions in the degradation of pollutant substances and improvement of sludge digestion in wastewater treatment (Chatel 2018). Furthermore, ultrasound provides reliability to new ideas in biotechnology and nanotechnology to engineering applications and drug delivery systems, medical imaging, and gene therapy (Liu et al. 2024a). The above applications support the general observation that ultrasound has become revolutionary across various fields, including the pharmaceutical industries and renewable energy.

One of the most substantial comparative benefits of ultrasound for chemical processes is that it can increase the rates of the processes (Sakthipandi et al. 2024). Because of its ability to enhance reaction rates, increase mass transfer, and produce highly reactive species, ultrasound provides results that other means cannot quickly obtain (Sancheti and Gogate 2017). Enhanced selective transformation occurs, energy consumption is lowest as localized energy effects arise, and resource utilization is least as the process is intensified (Shen et al. 2023a). These benefits correlate with contemporary chemical engineering goals, which aim to enhance processes while minimizing environmental and cost effects.

Still, there are a few issues that chemical engineers need to overcome to implement ultrasound in their work. Many technological challenges exist when large-scale laboratory processes are applied to industrial-scale reactors for energy efficiency. Ultrasound exposure has deleterious effects on several materials, demanding the use of durable building materials (Shi and Shi 2021). Moreover, the ability to control and predict the next stage of cavitation development in such systems is still considered a topic for further investigation (Zheng et al. 2022).

Ultrasonic technologies have a broad socio-economic impact (Jiang et al. 2020). Ultrasound can be an effective tool in industrial processes; it can control operational costs by using resources efficiently and minimizing wastage (Nabi et al. 2024). Its environmental friendliness also conforms to the current high standards set by the law for industries that need to meet these standards (Pham et al. 2013). Additionally, ultrasound is an environmentally friendly technology that supports the pursuit of localized and international environmental goals such as the United Nations Sustainable Development Goals (SDGs) (Obaideen et al. 2022). This underlines its function as an innovation hub and the catalyst for making the chemical engineering industry more sustainable.

The present chapter is designed to offer the reader a general introduction to the concept of ultrasonic technologies within the sphere of chemical engineering studies and practice and the potential directions for their further development. Further sections will provide more detailed information on general subjects, including sonochemical reactor design, its operation, reaction mechanisms, and the kind of equipment used; the modeling of the kinetic and mass transfer effects will also be discussed. Specifically, bubble dynamics, conditions that influence cavitation, and techniques for generating ultrasound will be discussed further, as well as measures for preventing contamination by air and the application of ultrasound in biological treatments of wastewater. The chapter will end with an understanding of energy control for energy-efficient chemical transformations as well as its impact on sustainability.

3.2 Sonochemical Reactors

Sonochemical reactors utilize ultrasonic cavitation action and the related energy to promote chemical reactions and ameliorate processes in chemical synthesis, environmental biotreatment, and pharmaceutical industries (Mason 2003). Ultrasonic waves, when they move through a liquid, create successively high-pressure and low-pressure areas that develop and cause tiny bubbles to grow and implode (Wu et al. 2021). This cavitation collapse creates very high-temperature hot spots up to 5000 K and pressure approximating 1000 atm that triggers reactions impossible under regular circumstances (Carlton 2019).

There are several types of sonochemical reactors, namely batch, flow, ultrasonic horn, ultrasonic bath, and hybrid reactors suitable for applications and scales (Manickam et al. 2023). These reactors can increase reaction rates, improve mass transfer, and raise the yield and selectivity of products. They are also useful in producing nanoparticles, emulsions, and pollutants (Adamou et al. 2024). The environment developed within sonochemical reactors is full of cavitation fields, providing several advantages for facilitating chemical transformations, many of them occurring with low-energy consumption and at the same time using small amounts of reagents and solvents (Suslick 2003). Recent developments in sonochemical reactor design and knowledge of the cavitation process have expanded the range of sonochemical processes and incorporated them more effectively into the industrial field.

3.2.1 Ultrasonic Horn Reactors

Ultrasonic horn reactors are among the most common sonochemical reactors, especially in the laboratory (Adamou et al. 2024). They use an ultrasonic horn placed into a liquid medium directly to control cavitation activity through high-frequency sounds (Janer et al. 2020). These reactors function at a fixed frequency with the cavitation intensity modulated by changes in power dissipation, usually in wave amplitude (Dehghani et al. 2023).

There are two basic designs: the so-called vertical horn, which creates high cavitation intensity for relatively low liquid volume (50–500 ml), and the horizontal horn, appropriate for higher capacities of the dispensing liquid (1–10 L). Although they are good at realizing local cavitation intensity, they are not very efficient at higher liquid volumes due to decreased power density (Adamou et al. 2024).

Batch operations make use of ultrasonic horns, although these can easily be incorporated into flow loops used in continuous processes. It covers the synthesis of chemical compounds and the oxidation process to reduce the size of particles and intensify mass transfer in multiphase systems (Adamou et al. 2024). Nevertheless, there are some weaknesses: contamination from the transducer erosion is often an issue in pharmaceutical and food industries, and dead zones—regions of lower cavitation intensity (Zupanc et al. 2019). These issues limit their applicability for large-scale production, although the use of several horns in a flow system can lessen these problems.

However, ultrasonic horn reactors are still convenient for local high-intensity applications in various small-scale processing (Pereira and Arantes 2018). They present a feasible approach for analyzing sonochemical impacts, capable of development and transition to more complex structures, such as multihorn systems, to handle concerns linked to cavitation and energy density homogeneity (Pokhrel et al. 2016) (Fig. 3.1).

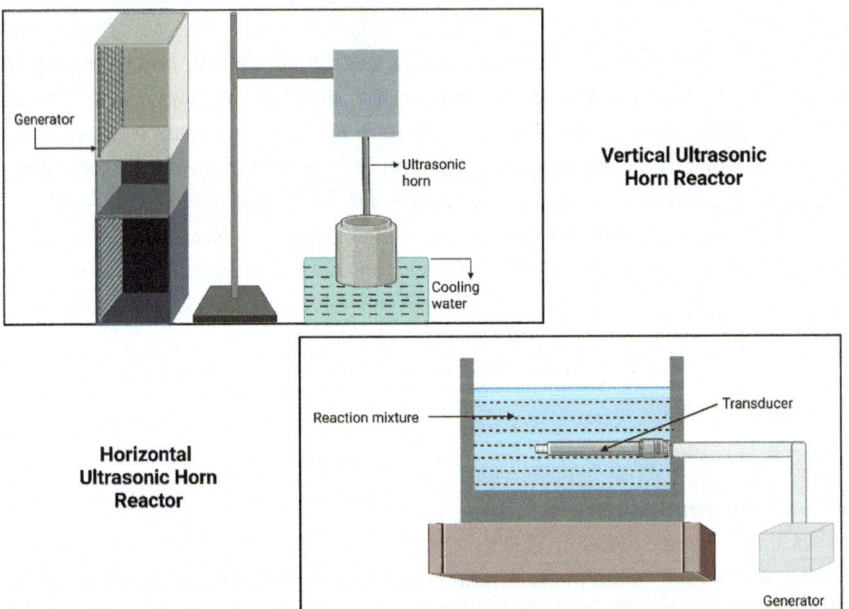

Fig. 3.1 Vertical ultrasonic horn reactors and horizontal ultrasonic horn reactors (created using Biorender.com)

3.2.2 Ultrasonic Bath Reactors

Ultrasonic bath reactors are typical in processes requiring a uniform cavitation activity spread over a larger base area (Dong et al. 2020). These reactors include a tank with several ultrasonic transducers at the bottom, producing ultrasonic waves through the coupling medium (Marhasin 2005). This means that the number of transducers can be variable depending on the required power density, while reactor capacity can be adjusted freely. Separately, these reactors have the possibility to work in batch or continuous production methods. Batch mode entails the direct discharge of the liquid into the bath, while in the continuous mode, clients use some means of flow, such as an overflow (Adamou et al. 2024). Moreover, ultrasonic baths can use direct or indirect irradiation; the first results in high levels of cavitation which are appropriate in such processes as water treatment, while the second employs a coupling fluid and is suitable in less rigorous processes such as enzymatic hydrolysis (Hoo et al. 2022; Verdini et al. 2024).

Ultrasonic baths can be designed as modular and have high cavitation activity; based on that, they are used in cleaning, emulsification, and chemical processes (Janer et al. 2020). The application of multiple transducers also guarantees the uniform distribution of cavitation pressure, providing better control over a range of processing conditions (Sivakumar et al. 2014). However, the two systems have disadvantages in that they experience energy losses at higher power levels. In the indirect irradiation system, the operating temperature is restricted more severely by factors such as the boiling point of coupling media like water (Asgharzadehahmadi et al. 2016). However, these difficulties have not stopped ultrasonic bath reactors from being stable and versatile vessels in laboratories and industries (Fig. 3.2).

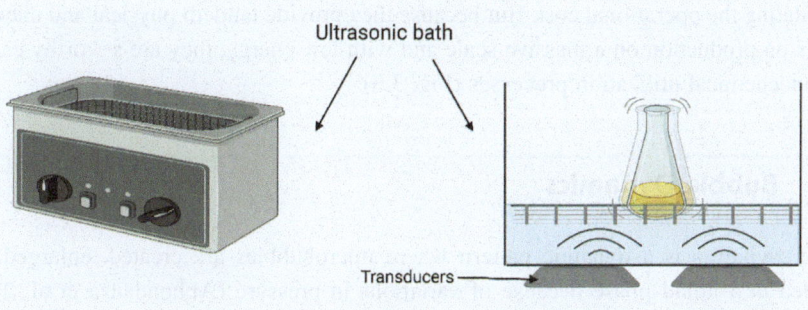

Fig. 3.2 Ultrasonic bath reactors (created using Biorender.com)

3.2.3 Multiple-Frequency Flow Cell Reactors

Flow cell reactors with multiple ultrasound frequencies are innovative sonochemical reactors intended for industrial-scale continuous processes. It involves mounting transducers on the outer or internal walls of the vessel, which can be arranged in rectangular or hexagonal patterns. These reactors employ the use of multiple frequencies to produce a physical action characterized by turbulence and liquid streaming for chemical action, including the formation of reactive radicals (Gogate and Patil 2016; Sancheti and Gogate 2017). For example, low frequency, for instance, 20 kHz, is used for physical effects and high frequency, for instance, 500 kHz, is used for chemical effects (Estivi et al. 2022). Flow cells, depending on the size of the required processing volume, can be produced from multiple units of a smaller size with identical cavitation activity, thanks to the modularity of the cell design (Dong et al. 2020).

The reactors are particularly applicable in cases where the process is continuous, such as chemical synthesis, wastewater treatment, and synthesis of nanoparticles (Banakar et al. 2022; Bao et al. 2023; Cristaldi et al. 2018; Dehghani et al. 2023). Therefore, through positioning and dosage of transducers, they guarantee optimal utilization of energy, as well as avoid issues such as transducer wear or the existence of lifeless regions in the area. Moreover, the length of the reactor itself can alter to adjust the time that the liquid is held in the reactor for, to meet certain tasks in the process. The integration of ultrasound with other technologies like UV irradiation has been proven to improve performance, particularly in advanced oxidation processes used in the degradation of pollutants (Lei et al. 2023).

However, multiple-frequency flow cells are somewhat more challenging to design and calibrate so that the cavitation intensity at the reactor remains uniform (Sancheti and Gogate 2017). A possible disadvantage of using multiple transducers and frequencies may be escalating the operational cost. But because they provide tandem physical and chemical impacts on production on a massive scale and with low energy, they are a worthy call for vast sonochemical utilization processes (Fig. 3.3).

3.3 Bubble Dynamics

Bubble cavitation is a dynamic pattern where microbubbles are created, enlarged, and imploded in a liquid phase because of variations in pressure (Abbondanza et al. 2023). In sonochemistry, pressure variations are caused by ultrasonic waves, which generally fall within the range of 20 kHz to a few MHz and travel through a liquid medium. Even as these waves move, they form an oscillating pressure construct of contractions and rarefaction pressure ridges, which in turn influence the manner in which different molecules in liquid behave (Ohl et al. 2015).

3.3 Bubble Dynamics

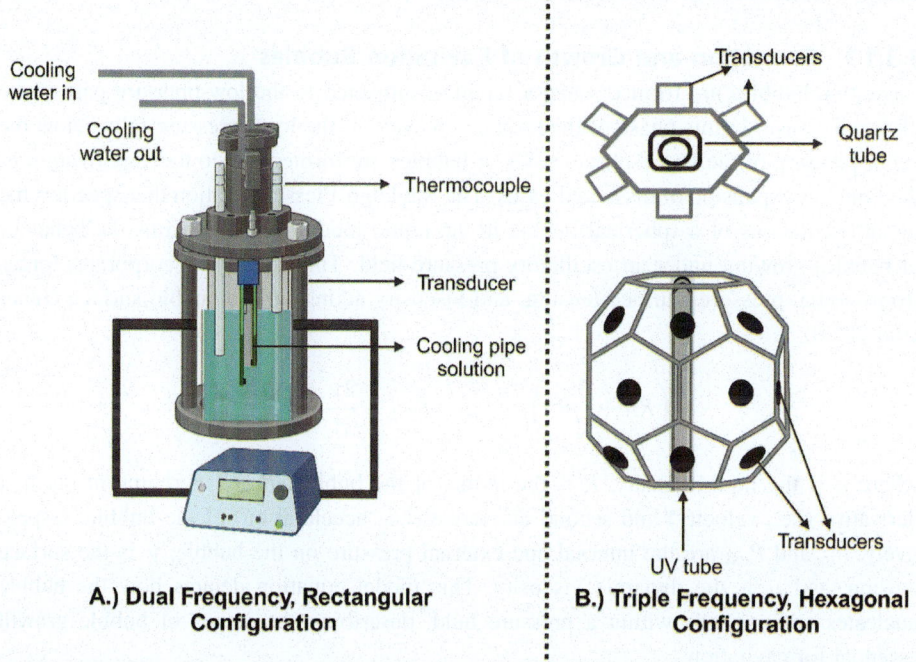

Fig. 3.3 Multiple-frequency flow cell reactors (created using Biorender.com)

During compressions, as the pressure increases, molecules are forced together. On the other hand, in the rarefaction phase, the pressure is low, and as a result, the molecules start to deflect, causing localized low-pressure form. In these circumstances, vapor bubbles may occur during nucleation and start growing. This cycle of pressure creates bubbles of alternating expansion and contraction, which, due to the expansion pressure enlarging the bubbles to an unstable size and then having them violently implode, is an explosive process. This rapid failure or cavitation event releases immense point energy within that liquid (Ashokkumar 2011).

The bursts result in hot spots with high pressure, meaning that copious amounts of energy are available for catalyzing reactions. These temperature and pressure hot spots can rise to the order of several thousand Kelvin and pressures, therefore providing local constraints necessary for reactions that would otherwise need very high or long periods of external conditions (Fattahi et al. 2024). The high pressure produced by bubble cavitation means that it is very useful in sonochemistry and can facilitate and enhance reactions in various fields.

3.3.1 Mechanism of Bubble Cavitation

3.3.1.1 Formation and Growth of Cavitation Bubbles

Cavitation bubbles are formed when a liquid is subjected to the low-pressure part of the ultrasonic wave. In this phase of the rarefaction cycle, if the local pressure falls below the vapor pressure of the liquid phase, voids or bubbles are formed due to the vaporization of the liquid or expansion of dissolved gases. The Rayleigh-Plesset equation that specifies the radial oscillations of a spherical bubble in the liquid medium best explains the behavior of bubbles growing under an oscillatory pressure field. This equation incorporates forces from inertia, pressure, surface tension, and viscosity acting on the bubble surface (Janer et al. 2020):

$$\rho\left(R\ddot{R} + \frac{3}{2}\dot{R}^2\right) = P_{in} - P_{ext} - \frac{2\gamma}{R} - \frac{4\mu\dot{R}}{R},$$

where ρ is the liquid density, R is the radius of the bubble, $\dot{R}$ and $\ddot{R}$ represent the first derivative (i.e., velocity) and second derivative (i.e., acceleration) of the bubble, respectively, P_{in} and P_{ext} are the internal and external pressure on the bubble, γ is the surface tension, and μ is the dynamic viscosity. This model equation depicts how the bubble nucleates and expands within a pressure field, describing the stages of bubble growth essential for cavitation.

3.3.1.2 Bubble Oscillation: Linear and Nonlinear Oscillations

As the bubble starts to expand, it starts to experience fluctuations in the pressure due to the oscillation of the ultrasound wave (Ashokkumar 2011). The periodic behavior of these variations—whether linear or nonlinear—solely depends on ultrasound waves "frequency" and "amplitude" (Baek et al. 2024). Linear behavior can be assumed at the low amplitude of oscillations when the bubble behaves as a simple linear oscillator. $R(t)$ changes harmonically, and its natural oscillation frequency can be expressed as:

$$\omega_0 = \sqrt{\frac{3}{\gamma R_0^3}},$$

where R_0 is the bubble's equilibrium radius. However, when oscillation amplitudes are being increased, nonlinear impacts start to manifest themselves. At high amplitude, the movement of the bubble no longer exhibits simple harmonic motion but complex behaviors like subharmonic, ultra-harmonic, and chaotic motion. Hence, these nonlinear oscillations indicate that the bubble is progressing toward instability and consequent collapse.

3.3.1.3 Collapse of the Bubble: Violent Implosion and Energy Release

In the collapse stage, the bubble collapses with tremendous speed and much force, generating extremely high internal pressures and temperatures (Yin et al. 2021). When the bubble has emerged to this size, it suddenly bursts and collapses violently to a very small radius. During this process of collapse, the surrounding liquid put high pressure on the bubble, creating localized high-energy "hot spots" within the bubble. The energy that is liberated during this collapse phase can also be determined by first evaluating the work that the liquid on the bubble performs as it contracts from the maximum bubble size R_{max} to its minimum radius R_{min} (Ashokkumar, 2011):

$$W = \int_{R_{min}}^{R_{max}} 4\pi R^2 P(R) dR.$$

If we assume the bubble collapse is an adiabatic process (no heat exchange due to the rapid timescale), the pressure inside the bubble $P(R)$ can be modeled as:

$$P(R) = P_0 \left(\frac{R_{max}}{R}\right)^{3\gamma},$$

where γ is the adiabatic index, and P_0 is the initial internal pressure. This compression leads to extremely high temperatures, which the relationship can estimate:

$$T = T_0 \left(\frac{R_0}{R_{min}}\right)^{3(\gamma-1)},$$

where T_0 is the initial temperature. This immense temperature, reaching thousands of Kelvin, creates conditions capable of initiating chemical reactions within the bubble and the surrounding liquid.

3.3.1.4 Rebound and Secondary Collapse

In order to explain the stranding process, which is the direct continuation of the first collapse, it is assumed that after the primary collapse, the bubble may go through many rebounds. Every succeeding re-expansion and collapse still has high-pressure and high-temperature conditions, yet not as severe as the first collapse. This "rebound effect" results in continuous cavitation activity that can be put to advantageous uses in the enhancement of mixing, maintenance of cavitation action or reaction conditions for greater periods, and desirable intensified chemical effects. These successive rebounds can be helpful in applications requiring maintaining a continuous cavity, such as nanoparticle synthesis, emulsification, or breaking environmental pollutants (Carlton 2019).

3.4 Factors Affecting Cavitation

Multiple factors, such as ultrasonic frequency, ultrasonic intensity, liquid properties, temperature, pressure, reactor design, and transducer location, affect the cavitation's efficiency and directly or indirectly affect the sonochemical output. Hence, it is paramount to study the factors which have been discussed below:

3.4.1 Ultrasonic Frequency

Ultrasound's frequency strongly impacts bubble behavior by changing the process of bubble formation, enlargement, vibration, and burst in sonochemical applications (Xiang et al. 2024). The acoustic frequency in determining the periods of the compression and rarefaction of cavitation bubbles controls their behavior and energy release (Husseini et al. 2005). Therefore, frequency selection is a key characteristic used in optimizing material response necessary to the desired physical and chemical effects (Saleh 2021).

For the bubble frequency levels with low oscillation rates ranging from 20 to 100 kHz, the rarefaction cycles are longer than the contraction cycles, enabling the bubbles to attain large sizes as they collapse (Wood et al. 2017). Such violent collapses produce pressure (up to 5000 bar) and temperature (up to 10,000 K) pulses, resulting in powerful local effects. Low-frequency ultrasound is preferred in terms of physical methods, which include the disintegration of particles, emulsification, and improved mixing brought about by shock waves and microstreaming owing to bubbles (Enomoto and Okitsu 2015). For instance, Son et al. showed that 35 kHz ultrasound was very efficient in the degradation of phenol in wastewater (Mahvi 2009). This increased the efficiency of breaking down complex organic pollutants since the intensity of the bubbles when they collapse is optimal. However, low-frequency cavitation has localized high intensity, reducing uniform energy distribution, mainly when used in large structures (Sharifishourabi et al. 2024).

High-frequency ultrasound using a 200–2000 kHz range creates tiny bubbles because of the short rarefaction cycle. These bubbles fall with less force but more often when they do so and produce localized hot spots that release the energy in small bursts. For this reason, high-frequency ultrasound is useful in chemical processes such as radical generation and selective synthesis (Janer et al. 2020). For instance, Lim et al. discovered that the 291 kHz ultrasound generated small and stable bubbles favorable to nucleation and growth of CdS nanoplatelets and minimizing agglomeration of particles with the attendant improvement of the product quality (Wang et al. 2003). Though high-frequency ultrasound is effective for chemistry applications, it may not offer the high mechanical energies needed in processes such as particle disruption or blending (Xiang et al. 2024).

Since there are apparent benefits and drawbacks to using each range, many studies have used dual-frequency or broadband systems in order to maximize the physical and chemical effects. For instance, wastewater treatment research used 40 kHz and 500 kHz frequencies, which made good use of the mechanical impacts of low frequencies and the chemical effects of high frequencies (Ghasemi et al. 2020). This approach enabled the achievement of the highest predicted pollutant removal by incorporating other cavitation behaviors such as sonochemical reactors. Such combined frequency systems show the possibility of efficient sonochemical processes for numerous applications (Zhang et al. 2015).

Several parameters describe the influence of the ultrasound frequency on bubble dynamics. As the frequency decreases, bubbles move bigger and implode in high powers in small bursts (Sims 1960). Small bubbles cease to implode violently at high frequencies but do so at a much higher rate, thereby providing continuous power output. This selection determines if the system creates physical activities such as turbulence and microstreaming or chemical activities such as free radical generation.

3.4.2 Ultrasonic Intensity

An ultrasonic intensity, the energy per unit area within the ultrasonic waves imposed upon a material, represents another essential parameter that directly influences the nature and characteristics of cavitation bubbles and the physics and chemistry of associated processes (Orthaber et al. 2020). The intensity determines the formation, stability, and destruction of cavitation bubbles, their effects and their contribution toward sonochemical efficiency.

At low ultrasonic intensities, the supplied energy is insufficient to support transient cavitation, a process involving the formation of bubbles and its violent collapse. However, stable cavitation is achieved, where bubbles oscillate around an equilibrium size and do not collapse disastrously (Carlton 2019). Ultrasonic pretreatment is beneficial for subsequent utilization of biomass, such as conversion of the biomass to valuable chemicals or materials, due to stable cavitation, which is gentle to the biomass and does not cause substantial damage, for example, in the disruption of biological cells or enzymatic reactions (Saif Ur Rehman et al. 2013). For example, in enzymatic hydrolysis tests, the low-intensity ultrasound enhanced the substrate-enzyme interaction without affecting enzyme structure, indicating the ecological conditions of mild cavitation for biochemical processes (Wang et al. 2018). However, low-intensity systems may not produce enough energy for applications that involve severe mechanical or chemical disturbance.

Low ultrasonic intensities induce stable cavitation where bubbles grow and selectively collapse with approximately one or two cycles, emitting great energy in sharply localized regions. High-intensity ultrasound is appropriate for emulsification, particle size reduction applications, and chemical synthesis. For instance, a study on ultrasonic treatment of wastewater showed that the ultrasonic intensity improved the degradation of organic and inorganic pollutants (Chen et al. 2024). Much higher rates of cavitation enhanced the generation of radicals, which promoted the degradation of tough-to-break molecules.

High-intensity ultrasound has strong therapeutic effects but no lack of problems. Overcoalescence of bubbles could occur at high intensities, and a range of possibilities is reduced, dampening the cavitation effect. Furthermore, energy losses that appear in heat diminish system efficiency and threaten to erode sensitive materials and parts of reactors. For instance, in the synthesis of nanoparticles, extremely large intensities resulted in linked particle formation and non-uniformity in particle size distribution, thus necessitating intensity optimization (Sreeharsha et al. 2024). In nanoparticle synthesis, moderate intensities indicate that the energy of bubble collapse is balanced with the controlled generation of radicals; the nucleation and growth are uniform (Chang et al. 2020). Likewise, in biodiesel production, the experiment revealed that with an intensity of 100–200 W/cm^2, the reaction rate and mass transfer improved, and therefore, the yields were comparatively higher than the conventional methods (Mohod et al. 2017). It is also important to note that the researchers can organize the system according to their needs by varying the ultrasonic intensity. At low intensity, more volume favors physical phenomena like microstreaming and better mass transfer, while at high intensity, more volume is favorable for chemical effects like bond breaking and the freeing of radicals (Khadhraoui et al. 2021). This versatility allows the ultrasonic systems to be applied in various fields, such as the food industry and water purification.

3.4.3 Liquid Properties

Cavitation efficiency is closely dependent on viscosity. In high-viscosity fluids, acoustic waves' amplitude is damped and bubbles' oscillation is limited; consequently, the level of cavitation is minimized (Bampouli et al. 2023). For instance, cavitation effects in polymer solutions are much less dominant compared to those in water because the viscosity of the medium hampers bubble activity (Gogate and Prajapat 2015). Low-viscosity fluids, on their part, allow for oscillation and subsequent collapse of bubbles and thus are suitable for most sonochemical uses such as water (Shen et al. 2023b).

It is evident that surface tension plays a crucial role in bubble nucleation and its stability. Smaller interfacial tension promotes bubble creation and stability, which would make cavitation more easily realized (Wang et al. 2022). Some other substances, like surfactants, are generally incorporated into the flow to achieve improved cavitation results. While studying the effect of ultrasonic emulsification, incorporating surfactants enhanced droplet size distribution and phase stability due to bubble stabilization (Merouani et al. 2024).

Vapor pressure dictates the amount of vapor inside a cavitation bubble and acts independently of the degree of cavitation (Zheng et al. 2022). Researchers found that fluid with moderate vapor pressure is most suitable because it does not allow bubbles to collapse at early stage but does not let them grow. High vapor pressure liquids, as in volatile organics, tend to reduce the strength and intensity of cavitation because the vapor in the bubbles expands excessively, leading to mild bubble collapse (Chowdhury and Viraraghavan 2009).

Supersaturated dissolves reduce the energy cavitation threshold by providing the site for the formation of bubbles (Scardina n.d.). Aerated systems provide better improvement of cavitation effects due to increased bubble numbers and the duration of their collapse (Han et al. 2024). For instance, the successful application of power ultrasound in wastewater treatment showed that systems with dissolved gases had greater degradation rates of pollutants than systems that had degassed conditions (Mahamuni and Adewuyi 2010). The effect of aeration on cavitation was to enhance the bubbles' growth and radicals as well.

During ultrasonic emulsification, the effect of adding surfactants to water was favorable for cavitation, giving better size distribution of the formed droplets and increased emulsion stability (Gaikwad and Pandit 2008). Likewise, "floating" in wastewater treatment systems appeared to enhance the degradation of organic pollutants in water by oxygen bubbles. Low-viscosity solvents such as ethanol in nanoparticle synthesis led to the simplest form of cavitation, hence the true size of particles with refined morphology (Pokhrel et al. 2016).

3.4.4 Temperature

At higher temperatures, the terminal velocity of oscillation of the cavitation bubble decreases, and hence, its collapse also becomes easier (Peng et al. 2020). Reduction in viscosity leads to improved passage of acoustic waves and greater cavitation intensity. Moreover, it decreases with the increase in temperature; in the rarefaction phase of ultrasounds, bubbles can form and grow more efficiently. Nevertheless, excessive temperatures have undesirable effects on cavitation (Olaya-Escobar et al. 2020). When temperature increases, vapor pressure also rises, causing an increase in vaporization in the bubbles (Speight 2019). This can reduce bubble collapse since the vapor acts as a cushion to reduce the amount of energy released. For instance, cavitation efficiency in wastewater treatment is observed to be high at an optimum temperature of 40–50 °C, while at high temperatures, the efficiency was observed to be low due to the high vapor pressure inside bubbles (Dular et al. 2016).

On the other hand, at low temperatures, the liquid viscosity and surface tension are high; therefore, bubble formation and oscillation are challenging to achieve (Mi et al. 2022). While this helps reduce the occurrence of cavitation, the collapses experienced are generally much more violent because of the lesser volume of vapor in the bubbles. For instance, experiments focused on free radical generation indicated that larger bubble collapses in lower temperatures could produce higher concentrations of radicals for oxidative reactions (Mark et al. 1998).

The sensitivity of cavitation efficiency upon temperature and its dependence on the application are stated. During nanoparticle synthesis, moderate temperatures of around 50 °C are used, as lower temperatures lead to higher viscosity and lower surface tension but could cause vaporization (Okoli et al. 2018). This leads to equal distributions of bubble collapse and precise control of nucleation and particle growth. In biodiesel production, reaction efficiency was also reported to decrease when working above 60 °C because of reduced mass transfer and bubble dynamics (Ghayal et al. 2013). After this point, the intensity of cavitation reduced since splashing inside the bubbles was over-vaporization.

In this respect, it should be noted that higher temperatures may adversely affect the system in some special conditions, such as enzyme-catalyzed processes. Even if cavitation is improved, enzymes may lose their activity at too high a temperature, which is common knowledge (Peterson et al. 2007). So, regulating temperature in such processes is essential to ensure the enzyme has optimal functionality while leveraging cavitation.

3.4.5 Pressure

One of the operational parameters that play a major role in cavitation phenomena is pressure, which impacts the nucleation, growth, and collapse of cavities. Thus, both hydrostatic and vapor pressures define the energy hurdles for cavitation and affect sonochemical efficacy (Zhang et al. 2023a).

Hydrostatic pressure has a way of working against cavitation in that at higher pressures, more energy is needed to form cavities (Vernès et al. 2020). Although this can reduce cavitation activity, the bubbles, which do form, suffer more violent collapses because of increased external pressure. This leads to increased energy loss and greater bubble temperatures, and pressures are produced during the collapse process. For instance, investigations concerning the effects of high pressure in wastewater treatment showed increased radical formation and degraded pollutants twice, owing to higher cavitation (Chakinala et al. 2008). Nonetheless, high pressures are characterized by lower incidences of cavitation events to the extent that applications involving steady cavitation events may experience reduced effectiveness.

Consequently, as the hydrostatic pressure is lower, the energy barrier for bubble formation is also reduced, which makes nucleation easier and results in more cavitation events (Theerthagiri et al. 2020). Nevertheless, the collapses are comparatively less sharp because the outer pressure does not build up sufficient compressive forces. This may lower the efficiency of applications involving high-energy bubble collapses such as chemical synthesis and nanoparticle formation. For example, a study on emulsification found out that increasing bubble numbers came with a cost of a weak shear forces that are important in reducing the droplet size (Hu et al. 2017).

Vapor pressure in the liquid is also taken into account when choosing the type of system to use. This is due to high vapor pressure, and more vapor is available to enter the bubble during the growth phase, thus minimizing the huff and puff effect and the energy released in the process (Zhang et al. 2023b). This is further facilitated at higher temperatures because vapor pressure is also high at such temperatures. Hydrostatic pressure is balanced to control the cavitation forces and their resultant behavior in such cases. Biodiesel production was investigated by comparing moderate pressures that enhanced cavitation efficiency based on the equilibrium between bubble stability and the collapse energy of bubbles (Crudo et al. 2016).

Pressure also influences the distribution and occurrence patterns of steady cavitation regions in the reactor. Hydrostatic pressures are high for such flow configurations, and they promote cavitation near the ultrasound source. Their impact makes high and concentrated cavitation zones. On the other hand, low pressure encourages more evenly distributed cavitation in the course of the liquid medium, and this might, at times, be convenient in enzymatic reactions or mixing.

In wastewater treatment, high pressures greatly enhance the breakdown of recalcitrant organic pollutants due to increased radical generation during bubble shrinkage (Jothinathan et al. 2021). On the other hand, moderate pressures were applied for food processing to minimize destructive effects on the tissue and cellular structure of the foodstuffs during cavitation (Arya et al. 2023).

3.4.6 Reactor Design and Transducer Placement

Cavitation relates to the formation of cavities in a flowing liquid, and the study is guided by the geometry of a reactor and the position of transducers. They determine the distribution of acoustic energy, the formation of active cavitation zones, and the efficiency of cavitation processes. Indeed, the position of the transducer and the geometry of the reactor are even more critical in large-scale or industrial system applications where the efficiency of the energy coupled and the effects to be produced are uniform (Moholkar 2015; Moholkar et al. 1999).

The reactor's orientation also dramatically influences the nature of the airborne ultrasonic waves in the liquid medium. Cylindrical reactors prove to be the most efficient for increasing the cavitation intensity due to resonance due to wave reflections from the surfaces of the reactor. This gives rise to the concentrated energy zones functional for such processes as chemical synthesis (Asakura et al. 2005). As an example of phenol degradation, the cavitation intensity index was increased using a cylindrical reactor that focused energy on specific areas, thereby increasing the reaction rate and pollutant degradation (Entezari et al. 2003). In contrast, the rectangular reactors or box-type designs offer a wider cavitation area that is optimal for processes that need uniform results, such as water purification (Adamou et al. 2024). As applied in continuously operated systems, the flow-through reactors employ extensions of geometries to realize work contact with the liquid and the acoustic field for an entire cavitation process path.

The position of transducers is as crucial here in controlling cavitation distribution as it is in other aspects of the reactors. Single-transducer systems, such as ultrasonic horns, are practical in applications requiring typical waves like nanoparticle production. These systems focused cavitation at the tip of the transducer, thus allowing reasonable control over particle formation and growth processes. However, they fail to produce uniform cavitation across large reactors and tend to form inactive or "dead zones" (Khairiyah et al. 2023). Issues of uneven energy distribution are solved in multitransducer systems where several transducers are used simultaneously. For example, a flow cell reactor in the shape of a rectangle with a total of six transducers in a hexagonal manner illustrated the improved cavitation uniform when used in biodiesel production for greater yield and reaction rates (Bargole et al. 2019). Transducers worked at multiple frequencies to further refine the nature of the pump and achieve a balance between physical and chemical effects.

The submersible transducers located in the liquid medium do not have energy loss by attenuation through the thickness of the reactor vessel. Such a configuration enables positive energy transfer, especially when baking, cooking, and using different industrial cleaning systems (Frank 1952). A case study in food processing revealed that direct immersion-type transducers improved cleaning performance and energy utilization by using submersible transducers to directly impart ultrasonic energy into the cleaning fluid (Sullivan 2009).

The observed patterning of the acoustic field by transducers in water leads to the formation of active and inactive cavitation zones. To appreciate the nature of the reactors' contents, one must appreciate that standing waves create highly active nodes in cylindrical reactors but leave anti-nodes relatively inactive. This localized intensity is advantageous for processes where specific effects are needed for a particular area of a system, such as chemical synthesis (Garcia-Vargas et al. 2023). On the other hand, the traveling wave distribution is relatively uniform and is most appropriate in big-scale systems. For example, wastewater treatment systems where transducers are installed at different reactor sections reduce dead zone areas, increase frequency cavitation, and enhance degradation rates (Dular et al. 2016).

3.5 Effects of Ultrasound on Chemical Systems

3.5.1 Solid–Liquid Systems

3.5.1.1 Ultrasonic Cleaning

Ultrasonic cleaning is widely adopted in electronics, pharmaceuticals, and other precision engineering industries. Microjets and high shear forces in the vicinity of surfaces are generated by collapsing cavitation bubbles in which unwanted substances such as oils, oxides, and particulates are dislodged (Fuchs 2015).

3.5 Effects of Ultrasound on Chemical Systems

In semiconductor manufacturing, silicon wafers are cleaned through ultrasonic frequencies greater than 1 MHz to remove submicron particles without deforming other smaller sensitive structures (Seike et al. 2024). This method produces a better result than chemical cleaning, using fewer chemicals and ensuring immaculate items. Using ultrasonic cleaners in washing lenses and mirrors removes dirt and remains without leaving behind abrasives, which may degrade the image-defining quality of equipment such as telescopes and cameras (Li and Khan 2016).

3.5.1.2 Extraction of Active Compounds

Ultrasound-assisted extraction (UAE) is a safe and effective technique for recovering bioactive compounds from plant origin (Kumar et al. 2020). Cavitation causes a breakdown of cell walls to improve permeation rates as well as the rate of diffusion (Zupanc et al. 2019). UAE is used in this study to recover polyphenols from grape seeds, giving a higher quantity than in conventional methods (Dzah et al. 2020). This application is essential in synthesizing antioxidants for developing nutraceuticals and cosmetic products. Plate transducers are sometimes known as large-scale ultrasonic extractors used for extracting essential oil from plants such as mint and hops (Richa et al. 2020). These systems also cut the extraction time from several hours to minutes while keeping the quality intact.

3.5.1.3 Membrane Filtration

Sonochemical processes enhance the lack of fouling and scaling in phenomena such as ultrafiltration and reverse osmosis on membrane surfaces (Qasim et al. 2018). They thought cavitation breaks up the layers of the deposit that make the linerless permeable. In wastewater treatment, ultrasound positively affects the filtration of total suspended solids and microbial load (Wen et al. 2024). Investigations reveal that this ultrasound method increases permeate flux by 40% when applied in cycles during filtration.

3.5.1.4 Dispersion of Nanoparticles

Ultrasound is essential to disperse nanoparticle nanoparticles such as titania, titanium oxide, and multiwalled carbon nanotubes in liquids (Sandhya et al. 2021). These resolve matters such as poor dispersion of reinforcements in a matrix by subjecting reinforcing particles to high shear forces resulting from cavitation, which is ideal for applications such as coatings and composites. Integrating single-walled carbon nanotubes by uniform dispersion using ultrasound improves polymer matrix-containing composites' mechanical and electrical properties (Zaib et al. 2019).

3.5.2 Liquid–Liquid Systems

3.5.2.1 Ultrasonic Emulsification

The emulsification applied by ultrasound is quite effective in forming relatively stable emulsions with small droplet sizes, even without adding surfactants (Kentish et al. 2008). In biodiesel production, ultrasonic emulsification combines vegetable oils and methanol to form microemulsion, tremendously accelerating reaction processes (Kojima and Takai 2019). They also reduce energy input while increasing biodiesel yield due to this method. In ultrasound-assisted homogeneous alkaline-catalyzed transesterification, alkaline catalysts such as sodium hydroxide (NaOH) or potassium hydroxide (KOH) are used in the reaction of triglycerides (TGs) with methanol to produce biodiesel and glycerol (Georgogianni et al. 2009). Ultrasonic irradiation makes it much more efficient and stimulates cavitation mixing, heat, and mass transfer. One of the major drawbacks of many of these traditional techniques is that oil and methanol do not mix easily, and this technique handles this problem very effectively.

The main reaction involves the stepwise conversion of triglycerides into fatty acid methyl esters (FAMEs) and glycerol. The overall reaction can be summarized as:

$$\text{Triglycerides} + 3\,\text{MeOH} \xrightarrow{\text{Catalyst}} \text{Gylcerol} + 3\,\text{FAME(biodiesel)}.$$

Due to ultrasonication, reaction kinetics are accelerated, and the activation energy is reduced. Furthermore, the surface contact between methanol and oil is enhanced by forming an emulsion, thus improving the transesterification reaction rate.

A notable side reaction is saponification, where free fatty acids (FFAs) react with the alkali catalyst to form soap and water:

$$\text{FFA} + \text{NaOH} \rightarrow \text{Soap (RCOONa)} + \text{H}_2\text{O}.$$

This reaction reduces biodiesel yield by consuming the catalyst, but ultrasound minimizes this problem by enhancing the rate of transesterification, thus allowing little time for side reactions. Ultrasound enhances and speeds up reaction time (30 min as against hours in other methods) and has low alcohol-to-oil molar ratios, small catalyst amounts, and no external heating due to heat produced by cavitation effects. Apiculate transesterification processes also ease the separation of biodiesel from glycerol through ultrasound radiation, which lowers downstream costs. Lotions and creams produced utilizing ultrasound have improved the texture and preservation of micro- and nanoparticles of vitamins and coenzymes (Kaci et al. 2018).

3.5.2.2 Nanoemulsion Formations

Nanoemulsions were prepared by subjecting the samples to ultrasonication at different frequencies, resulting in a droplet size of less than 200 nm. Nanoemulsions prepared using an ultrasonic technique are employed in the delivery of hydrophobic drugs such as curcumin, where solubility and targeting are enhanced (Modarres-Gheisari et al. 2019). Ultrasonic technology used here has been found to strengthen aspirin nanoemulsion drug delivery, owing to the effects of ultrasound cavitation, especially on the stability and efficiency of drug formulations. The ultrasonic process creates highly concentrated energy to directly disrupt droplets into nanoemulsions, which offer a large surface area for aspirin dissolution. The formation of smaller droplets at the nanoscale increases solubility and provides a better means of delivering the drug (Preeti et al. 2023).

Ultrasound technology helps maintain a constant droplet size, which is important for the stability and performance of any nanoemulsion (Fathordoobady et al. 2021). These formulations are anti-creaming, anti-flocculation, and anti-settling in nature, for the most part. Furthermore, the technology enables viable encapsulation of aspirin in oil in water (O/W) nanoemulsion and water in oil in water (W/O/W) nanomultiple emulsions (Tang et al. 2012). This encapsulation minimizes the interaction of the aspirin with the gastrointestinal lining, thus decreasing side effects such as gastrointestinal irritation. In addition, W/O/W emulsions, obtained using two-stage ultrasonic cavitation, also possess the functionality of prolonged and/or continuation of drug release due to the multiple layering of the emulsion system (Mahmood et al. 2014).

The preparation of nanoemulsions through ultrasound leads to an improvement in the therapeutic benefit. Research established that these formulations cut inflammation and pain much more than traditional aspirin suspensions. A study using animal models showed that nanoemulsions significantly reduced paw edema caused by carrageenan, thus proving to be decisive in fighting inflammation (Marwa et al. 2023). Further, it is recorded from acetic acid-induced writhing as well as hot plate models that the nanoemulsions resulted in better pain control over standard aspirin suspension (Tang et al. 2012).

Another advantage of ultrasonic technology in the formation of nanoemulsions is that it can incorporate some penetrants, such as transcutol (Tanuku et al. 2024). It also enhances the therapeutic performance of aspirin by helping the drug to pass through biological membranes through these components. Besides, the nanoemulsion system of encapsulation introduced here has the added advantage of sparing the gastrointestinal tract from irritation, which is characteristic of aspirin tablets. Such systems improve the overall effectiveness of the therapy and minimize adverse reactions in case of cancer.

3.5.3 Gas–Liquid Systems

3.5.3.1 Ultrasonic Atomization

Positive pressure atomization works by shattering the surface of the liquid into droplets by using high-frequency sound waves known as ultrasonic waves; hence, it is used in humidification, spray drying, and aerosols. Fermentation ultrasonic atomizers minimize the number of water droplets that concentrate the ethanol level of products such as sake.

3.5.4 Reaction Processes in Chemical Engineering

3.5.4.1 Polymerization

Sonochemical polymerization improves polymer yield through radical production during cavitation (Santha Kumar et al. 2023). To produce latex particles with a narrow particle size distribution that can be used in paints and coatings, styrene is polymerized under sonic frequencies (Kalita et al. 2024). Sonochemical techniques are used to control the polymerization of monomers such as ethylene and vinyl acetate for synthetically developing polymers for adhesives and films (Gogate and Prajapat 2015).

3.5.4.2 Heterogeneous Catalysis

Ultrasound improves catalytic processes by distributing catalysts evenly and raising their effective surface. In the hydrogenation of vegetable oil, the application of ultrasound promotes high reaction rates and uniform use of the catalyst to shorten the process timeline (Pukale et al. 2015).

3.5.4.3 Decomposition of Pollutants

Ultrasound is effective in the process of oxidative destruction of organic compounds, including dyes and pesticide residues, in wastewater (Wang et al. 2023). Hydroxyl radicals produced by ultrasound of a frequency of 40 kHz degrade methyl orange dye in water with a removal efficiency of 95% in 30 min (He et al. 2016). Some of the ultrasonic processes in environmental remedial applications exclude using toxic chemical reagents, an aspect of green chemistry (Wu et al. 2013).

3.5.4.4 Enzymatic Reactions

Ultrasound assists in increasing the activity of enzymes by increasing the adhesive on the substrate and hence increasing the rate of diffusion limitations (Wang et al. 2018). In biodiesel production, ultrasonication enhances lipase's action, causing an increased conversion of triglycerides to fatty acid methyl esters (Bhangu et al. 2017).

An instance of how ultrasound influences the process is the esterification of oleic acid, a long-chain fatty acid ($C_{18}H_{34}O_2$) with triethanolamine $C_6H_{15}NO_3$ catalyzed by lipase to a diester (Masoumi et al. 2011). The conventional method takes 8 h at 160 °C, and enzymatic esterification at room temperature takes 24 h. Ultrasound-assisted enzymatic esterification, on the other hand, takes only 2.5 h at 40 °C with more than 94% of esterification yield. The enhancement of the reaction rates can be explained by the ability of ultrasound to generate microstreaming and shock waves, which increase the molecular collision; the ability to increase the diffusion of the substrate to the active site of the enzyme; and the mechanical effect provision of mechanical energy to encourage the performance of enzymes.

Mathematically, the enhancement in reaction rate due to ultrasound is represented by the order: $k_{EUS} > k_{ENUS} > k_{NUS}$. This can be explained using the Arrhenius equation, which describes the reaction rate constant k as $k = Ae^{-\frac{E_a}{RT}}$, where A is the frequency factor, E_a is the activation energy, R is the gas constant, and T is the temperature. Ultrasound reduces the effective activation energy E_a, leading to faster reaction rates.

k_{EUS}: Rate constant of the ultrasound-assisted enzymatic reaction.

k_{ENUS}: Rate constant of the enzymatic reaction at room temperature.

k_{NUS}: Rate constant of the conventional reaction method.

3.6 Modeling of Mass Transfer Effect

3.6.1 Sonochemical Factors Affecting Mass Transfer

The mass transfer coefficient characterizes the situation with mass transfer in sonochemical reactors. K_L proves to depend on the operational and physical parameters defining the cavitation processes. The capability of utilizing ultrasound power is directly associated with K_L through enhancing the energy density of the cavitation zones (Sajjadi et al. 2017). Higher power increases the collapse intensity of the bubble and the turbulence, which results in better interfacial renewal and a higher K_L (Hoo et al. 2022). However, when the power level is increased beyond this optimum level, coalescence of bubbles occurs, and there are fewer active cavitation sites, resulting again in a low K_L (Yamashita and Ando 2019). The frequency of ultrasound also plays a vital role: low-frequency ultrasound (20–40 kHz) produces larger and more energetic bubbles, leading to better turbulence and interface renewal and a greatly enhanced K_L. Conversely, high-frequency ultrasound (500 kHz–1 MHZ) creates less energetic bubbles, smaller in size with lesser effect on the K_L hence uniformity in cavitation.

The specifics of the liquid, including viscosity, surface tension, and diffusivity, influence K_L significantly (Song et al. 2017). The observed turbulence of low-viscosity liquids enables bubbles to oscillate and collapse, enhancing K_L. Conversely, the higher viscosity mitigates these effects and reduces K_L. Likewise, since a higher surface tension of a liquid reduces bubble formation, K_L goes up since relatively more low surface tension fluids form bubbles more easily, allowing more harsh collapses and more mass transfer—which tends to increase K_L (Boubendir et al. 2020). Additional components such as electrolytes and surfactants can further enhance K_L based on bubble stability, their numbers, and the area of the interface available. Conversely, K_L decreases though cavitation intensity will also fall as concentration of surfactant increases; it was seen as a diminished parameter (Merouani et al. 2024). Again, the rates of gas flow affect K_L, where moderate flow rates promote bubble, interfacial activity, and consequent increases in K_L. But turbulent flow assists coalescence, thus decreasing the number of active cavitation zones and K_L.

Other factors that affect K_L include reactor geometry together with operating temperature (Ramezani et al. 2015). Ultrasonic horns mainly have high-concentration cavitation zones and, hence, higher values of K_L while ultrasonic baths give general weak cavitation and, thus, moderate enhancement of K_L (Adamou et al. 2024). A more appropriate larger location enhances the correspondence of cavitation regions with K_L and gas–liquid interfaces. In general, as temperature rises, K_L increases due to a decrease in liquid viscosity and in turn an increase in liquid diffusivity; however, at excessively high temperature, intensity of cavitation is affected negatively lowering K_L (Caruso et al. 2009). Additionally, bubble size and dynamics are critical. The design also argued that the formation and the reduction of bubble size enhance the interfacial area to enhance K_L, and the larger bubbles increase turbulence, which is also beneficial to K_L (Feng et al. 2019). All of these factors explain why so much attention should be paid to the optimal control of the mass transfer efficiency in sonochemical reactors.

3.6.2 Transfer Correlations for Ultrasonic Reactors

The volumetric mass transfer coefficient (K_La) in ultrasonic horn reactors can be correlated with a power density $\left(\frac{P}{V}\right)$ and gas flow rate $\left(V_g\right)$ as(Kumar et al. 2004):

$$K_La = 0.029 \left(\frac{P}{V}\right)^{0.17} V_g^{0.37}.$$

It was also identified from the experimental studies that K_La can be improved by 50–110% by using ultrasound when compared to those systems that use mechanical agitation (Hong et al. 2023). The mass transfer coefficients at the vicinity of the horn tip are far higher than those in other sections of the reactor due to the high cavitation intensity (Faïd et al. 1998). These reactors are particularly useful for processes where high-intensity mass transfer is needed in a small part of the reactor, such as nanoparticle production or only one reaction in a single-phase system.

Ultrasonic baths split energy over the largest volume, providing a more extensive but not so powerful cavitation field than ultrasonic horns do. The volumetric mass transfer coefficient (K_La) for ultrasonic bath reactors is presented as follows (Kumar et al. 2004):

$$K_La = 0.0039 \left(\frac{P}{V}\right)^{0.4} V_g^{0.6}.$$

Horn reactors are characterized by localized cavitation intensities at the horn tips, which cause a higher mass transfer coefficient K_La due to the extreme bubble collapse and turbulence around the horn tip. On the other hand, ultrasonic bath reactors energize a larger volume uniformly but with lower energy density, therefore yielding moderate K_La results (Faïd et al. 1998). Thus, horn reactors are preferable for localized and highly concentrated energy input as for ultrasonic baths, whereas bulk treatment requiring uniform energy distribution is preferable. Further, horn reactors are less sensitive to the power density and gas flow rate than bath reactors because energy in horn reactors is delivered

3.6 Modeling of Mass Transfer Effect

in a concentrated manner. This makes bath reactors transport dependent and variations in such parameters as power density, and gas flow rate have a profound impact on mass transfer rate (Kumar et al. 2004).

3.6.3 Diffusion-Limited Model for Vapor Transport in Cavitation Bubbles

According to the diffusion-limited model, the vapor transport through the oscillating cavitation bubbles is governed by the assumption of Storey and Szeri that vapor transport and entrapment are fundamental diffusion processes (Storey and Szeri, 2001). To extend the model, Toegel et al. proposed a new model based on the ODEs to represent the diffusion of the vapor molecules within the bubble (Masiello et al. 2024). This model employs the boundary layer approximation and has been compared to the PDE mode as well as a simplified model of Storey and Szeri.

The diffusion-limited model takes into account certain fundamental physical and thermodynamic processes, such as bubble dynamics, heat conduction, and mass transfer across the bubble interface. The model relies on four key components (Moholkar 2015):

3.6.3.1 Mathematical Model in the Form of the Keller-Miksis Equation for the Radial Motion of Cavitation Bubbles

The radius of the bubble is given by:

$$\left(1 - \frac{dR/dt}{c}\right) R \frac{d^2 R}{dt^2} + \frac{3}{2}\left(1 - \frac{dR/dt}{3c}\right)\left(\frac{dR}{dt}\right)^2$$
$$= \frac{1}{\rho_L}\left(1 - \frac{dR/dt}{c}\right)(P_i - P_t) + \frac{R}{\rho_L c}\frac{dP_i}{dt} - 4\nu\frac{dR/dt}{R} - \frac{2\sigma}{\rho_L R}$$

where $t = 0$ & $R = R_0$.

The bubble wall velocity can be induced from internal pressure in the bubble.

$$P_i = \frac{N_{\text{tot}}(t)kT}{\left(\frac{4\pi(R^3(t)-h^3)}{3}\right)} \quad \text{where} \quad \frac{dR}{dt} = 0.$$

3.6.3.2 Diffusive Flux Equation for Solvent Vapor Transport Through the Bubble Wall

The number of solvent molecules in the bubble (N_w) is given by

$$\frac{dN_W}{dt} = 4\pi R^2 D_W \left.\frac{\partial C_W}{\partial r}\right|_{r=R} \approx 4\pi R^2 D_W \left(\frac{C_{WR} - C_W}{l_{\text{diff}}}\right).$$

The above equation holds good for bubble in motion where bubble velocity $\left(\frac{dR}{dt}\right)$ is nonzero. Hence, we use vapor transport inside the bubble which becomes a pure diffusion equation.

$$\frac{\partial C_W}{\partial t} = D_{ij}\left(\frac{\partial^2 C_W}{\partial r^2} + \frac{2}{r}\frac{\partial C_W}{\partial r}\right).$$

Applying boundary equations for the above equation are:

$$r = 0, \frac{\partial C_W}{\partial t} = 0 \text{ for } t \geq 0$$

$$r = R, C_W = C_{WR} \text{ for } t \geq 0$$

$$C_W = C_{W0} = 0 \text{ for } t \geq 0 \; 0 \leq r \leq R.$$

The analytical solution is then will be:

$$\frac{C_W - C_{W0}}{C_{WR} - C_{W0}} = 1 + \frac{2R}{\pi r}\sum_{i=0}^{\infty}\frac{(-1)^n}{n}\sin\left(\frac{nr}{R/\pi}\right)\exp\left(-\frac{n^2 D_{ij} t}{(R/\pi)^2}\right).$$

Here, the instantaneous diffusive penetration length is given as follows:

$$l_{\text{diff}} = \min\left(\sqrt{\frac{RD_S}{\left|\frac{dR}{dt}\right|}}, \frac{R}{\pi}\right) \text{ where } N_W = 0.$$

3.6.3.3 Heat Conduction Equation Through the Bubble Wall

$$\frac{dQ}{dt} = 4\pi R^2 \lambda \left.\frac{\partial T}{\partial r}\right|_{r=R} \approx 4\pi R^2 \lambda \left(\frac{T_0 - T}{l_{\text{th}}}\right).$$

Here, the thermal diffusion length is given as follows:

$$l_{\text{th}} = \min\left(\sqrt{\frac{R\kappa}{\left|\frac{dR}{dt}\right|}}, \frac{R}{\pi}\right) \text{ where } Q = 0$$

3.6.3.4 Energy Balance Equations on Cavitation Bubble Dynamics in Submerged Liquid Flow

The following equation gives the energy balance

$$C_{V,\text{nix}}\frac{dT}{dt} = \frac{dQ}{dt} - P_i\frac{dV}{dt} + (h_W - U_W)\frac{dN_W}{dt}$$

Mixture heat capacity: $C_{V,\text{nix}} = \sum C_{V,i} N_i$

Enthalpy: $h_W = \left(1 + \dfrac{f_i}{2}\right) kT_0$

Internal Energy: $U_W = N_W kT \left(3 + \sum_{i=1}^{3} \dfrac{\theta_i/T}{\exp(\theta_i/T) - 1}\right)$

Heat capacities of other species ($i = O_2, N_2, H_2O$)

$$C_{V,i} = N_i k \left(\dfrac{f_i}{2} + \sum \left(\dfrac{(\theta_i/T)^2 \exp(\theta_i/T)}{(\exp(\theta_i/T) - 1)^2}\right)\right).$$

The transport parameters envisaged with heat and mass transfer are computed by Chapman-Enskog theory with Lennard–Jones 12–6 potential as a foundation for determining the diffusion coefficient of the vapor molecules in the liquid. These parameters are then used to compute thermal and diffusive penetration depths from the radial motion of cavitation bubbles.

3.7 Methods Used to Produce Ultrasound

The production of ultrasound in chemical processes involves several methods, each utilizing different technologies to generate high-frequency sound waves that induce cavitation and improve reaction rates (Xie et al. 2024). Below are the primary methods used to produce ultrasound:

3.7.1 Piezoelectric Transducers

Leading candidates for medical imaging applications include PZT-5H (lead zirconate titanate) ceramics. These materials have high electromechanical coupling (k33 > 75%), corresponding to good bandwidth and sensitivity for ultrasound production in the 3–7 MHz frequency range. For higher frequencies, materials like polycrystalline lead magnesium niobate-lead titanate (PMN-PT) represent improved electrical impedance matching and also more substantial coupling coefficients compared to lead zirconate titanate, ensuring high-resolution applications in the 20–100 MHz range (Zhang et al. 2024).

Single Crystal Relaxor-PTs: Advances in single crystal Relaxor-PTs, which have the ability to provide coupling coefficients more significant than 90%, enable high bandwidth ultrasound transducers. These materials are applied in advanced diagnostic techniques like contrast and harmonic imaging, primarily when they aim at applications requiring high resolution for small animal and ophthalmic imaging (Li 2024).

Fabrication Techniques: Various fabrication methods are applied to create piezoelectric transducers. Such methods include: (Sharma et al. 2024)

Dice and Fill Method: Cutting piezoelectric materials into small pieces and then filling the gaps with epoxy are based on this method. It is generally applied for making 2–2 composite arrays (Li et al. 2024).

Tape Casting and Sol–Gel Methods: These are alternative methods used to fabricate high-frequency transducers (20–40 MHz). These techniques enable the realization of fine-scale piezoelectric elements (Zhao et al. 2024).

Deep Reactive Ion Etching: DRIE is an advanced technology that allows for microscale features in critical applications. High-frequency ultrasound transducers requiring frequencies above 40 MHz require such a level of precision. It makes the exact patterns down to fine and accurate features necessary in ultrasound applications with high-resolution requirements (Laermer et al. 2020).

3.7.2 Magnetostrictive Transducers

Magnetostrictive transducers generate ultrasound by utilizing magnetic materials that deform under the influence of a magnetic field. When an electric current is applied, the material undergoes mechanical deformation, producing high-frequency sound waves. This method is effective for generating ultrasonic waves, particularly for applications requiring precise control over ultrasound frequency and intensity (Devos et al. 2025).

3.7.3 Ultrasonic Horns or Probes

These are employed to amplify the ultrasound waves into a medium and direct them to a specific area. This increases the cavitation effects by focusing the sound waves into the target area, thus increasing energy intensity in emulsification or extraction (Hofmann et al. 2023).

3.7.4 Laser-Generated Ultrasound

This technique employs short pulses from the laser to generate ultrasound waves by rapidly heating a target material. The rapid expansion of the material creates acoustic waves, which are useful in precision applications (Torrisi et al. 2024).

3.7.5 Electrostatic or Electroacoustic Methods

Ultrasound can also be generated by applying an electric field to specific materials, causing them to vibrate and emit sound waves (Rennoll et al. 2023).

3.8 The Control of Airborne Contamination

Airborne contamination is highly relevant to chemical engineers because this activity impacts public health, environmental quality, and industrial processes. Airborne pollutants include particulate matter (PM), trace elements, and microplastics. The sources may be fossil fuel combustion and industrial and urban activities such as construction and recycling of waste. The contaminants lead to a range of diseases, from respiratory ailments to cancer, where regions in South and Southeast Asia are seeing increased harmful elements like lead and arsenic due to e-waste recycling and coal-fired kilns, as well as through efforts in emission monitoring and reduction, such as the one implemented in the SPARTAN network and targeted intervention (Yuming et al. 2022). Ultrasonic agglomeration, a cutting-edge method for air filtration enhancement, relies on standing wave fields to facilitate increased particle collisions, forming larger agglomerates than moving wave fields. This process is implemented in a specialized aerosol wind tunnel equipped with a 21 kHz ultrasonic transducer, effectively mixing aerosolized particles with ambient air. Cyclone separators are then employed to capture these larger agglomerates, reducing downstream particle mass and minimizing pressure drop across filters. This approach proves highly efficient by extending filter lifespan by over 50% and lowering energy demands in ventilation systems. Key factors like residence time, sound pressure levels, and particle concentration are carefully optimized, making ultrasonic agglomeration a sustainable solution for managing airborne contamination in industrial systems (Liu et al. 2024b).

In the paper, Somanathan et al. propose a novel route for synthesizing metal oxide nanoparticles directly from industrial waste streams. CeO_2 was used for the first time in photocatalytic degradation of indoor air pollutants. Materials were synthesized by hydrothermal and anaerobic processes involving electroplating sludge and iron ore tailings-Fe_3O_4. The electrochemical systems used catalytic activity to degrade pollutants like acetaldehyde and methylene blue by UV light and H_2O_2, evaluated through linear sweep voltammetry. Antibacterial efficacy of Fe_3O_4 and Ag–Fe_2O_4 nanomaterials compared to hydrophobic Ag and Fe_2O_4 showed their potential in pollution control (Somanathan et al. 2024). Ultrasound energy greatly enhances nucleation and significantly reduces the duration of crystallization. Thus, sonochemistry has emerged as an emergent technology that employs ultrasonic power to synthesize metal–organic frameworks (MOFs) (Chatterjee et al. 2024).

S. Upendran, E. K. Jyothi, A. Jose, and R. G. Nath developed the Portable UVC Fan Coil Unit, or FCU, system, utilizing UVGI technology to significantly reduce indoor microbial contamination by targeting airborne viruses and bacteria. Experimental studies have demonstrated its effectiveness, with a 97% reduction in bacterial colony-forming units within 30 min of operation and up to 98.5% at 90 min. Consistent results were obtained over many days, which establishes its reliability. Real-time air sampling was used to determine microbial counts. The system is fitted with high-output UVC lamps and prefilters that remove larger-sized particles, allowing safe operation without human exposure to direct UVC radiation. Validated for airborne diseases such as SARS-CoV-2 and Influenza, the UVC-FCU destroys protein capsids and lipid envelopes. Therefore, this device is a very effective tool in improving indoor air quality (Upendran and Jose 2022).

The topic of airborne microplastics and their potential health impacts has gained significant interest. There have been suggestions that such small particles might be transported long distances through the air, thereby contributing to plastic pollution in the atmosphere. Airborne microplastic contamination has been less emphasized than plastic pollution on land and in the sea. Businesses involved in textile or apparel production generate large quantities of these particles. Research indicated that wind-driven movement is responsible for an estimated 7% of the microplastics that enter the ocean. Particles of microplastics in the atmosphere may reach 750 µm in length, with daily deposition rates in Germany ranging from 137 to 512 particles per square meter. Additionally, Dris et al. found that interior environments have more concentrated microplastic quantities with notable day-to-day settling rates. The health hazards associated with airborne microplastic exposure form an essential impetus to install effective mechanisms for monitoring and controlling it (Kwon et al. 2023).

Table 3.1 summarizes the various methods for indoor air quality, including air filtration systems like viscous impingement filters and electrostatic precipitators, gas and vapor removal with high-tech filtration, microbial control using UV germicidal lamps, and ventilation, humidity, and temperature regulation are given. Regular maintenance is essential to ensure effective performance.

3.9 Wastewater Treatments

Ultrasound technology is increasingly being utilized in biological wastewater treatment due to its ability to enhance the degradation of organic pollutants and improve the efficiency of biological processes. Biological wastewater treatment is one of the practical solutions to meet the increasing demand for appropriate methods addressing the increased generation of wastewater, as it uses microorganisms to degrade organic pollutants. Low-intensity ultrasound irradiation (LUSI) improves treatment efficiency by disrupting microbial cells, boosting enzyme activity, and improving mass transfer via cavitation effects. LUSI benefits aerobic and anaerobic processes by increasing Chemical

Table 3.1 Prevention of airborne contamination given by Aslam et al. (2024)

Category	Method/Actions	Details	References
Air filtration systems	Viscous impingement filters	Traps particles on oil-grease-coated surfaces; available as reusable, cleanable, or disposable types	Agarwal et al. (2020)
	Dry screen filters	Relies on fine mesh screens without adhesives to filter out contaminants	Kamøy et al. (2022)
	Electrostatic precipitators	Electrically charges particles; oppositely charged plates remove particles; cleaned with hot water	Afshari et al. (2020)
	Air washers	Poor efficiency; removes about 60% of airborne contaminants by weight	Omelchenko et al. (2022)
Gas and vapor removal	High technology filtration systems	Uses activated carbon, potassium permanganate, and other catalysts to neutralize gaseous pollutants	Zhu et al. (2022)
Microbial control	UV germicidal lamps	Employs ultraviolet light to inactivate airborne bacteria and viruses	Goswami and Pugazhenthi (2020)
Ventilation and humidity control	Humidity control	Maintains relative humidity between 30 and 70% to limit microbial growth and ensure comfort	Gilmanova et al. (2021)
	Natural or mechanical ventilation	Facilitates fresh air exchange through ducts, windows, doors, or structural openings. https://doi.org/10.1016/j.buildenv.2023.110726	Torresin et al. (2023)
	CO_2 level monitoring	Keeps carbon dioxide levels below 700 ppm above the outdoor baseline	Kormušoski et al. (2024)

(continued)

Table 3.1 (continued)

Category	Method/Actions	Details	References
Temperature management	Regulated indoor temperatures	Maintains summer temperatures at 21–22.5 °C and winter temperatures at 18.5–21 °C for comfort and health	Zhou et al. (2022)
Scheduled maintenance	Regular filter cleaning or replacement	Ensures timely maintenance of filters to prevent reintroduction of pollutants	Li and Siegel (2021)

Oxygen Demand (COD) and nitrogen removal, reducing sludge production, and enhancing methane yields in anaerobic digestion. However, potential microbial damage, reduced sludge settleability, and optimization of operational parameters present the primary challenges. Future research in the field should focus on real-world applications, parameter optimization, and LUSI in connection with other technologies to maximize its benefits (Zhang and Jin 2015).

In biological wastewater treatment, gases such as CO_2, CH_4, N_2, and SO_2 are emitted when converting suspended and dissolved organic contaminants to biomass. The activated sludge process, while effective—it can treat wastewater up to 10 times more efficiently per reactor volume than any other approach—is characterized by the production of surplus sludge that constitutes 50–60% of the treatment costs. Techniques such as ultrasonic waves for sludge disintegration and lysis-cryptic growth, where the lysed cells release nutrients for microbial reuse, have been studied to decrease sludge production. Research on ultrasonic treatment in sequencing batch reactors (SBRs) found that several parameters influence sludge reduction, including power, exposure period, and specific supplied energy. An ideal Specific Sludge Energy (SSE) of 35,000 kJ/kg Volatile Suspended Solids (VSSs) can reduce up to 78% sludge while meeting discharge regulations. High power use, however, has a disastrous impact on COD removal and effluent quality. Advanced oxidation, biological hydrolysis, and mechanical, chemical, and thermal lysis are other sludge reduction methods. Ultrasonic waves enhance sludge sedimentation and reduce microbial activity, so energy input and operating parameters must be effectively optimized (Mohammadi et al. 2011).

The process significantly enhances biological wastewater treatment mainly through ultrasound technology's mechanical, thermal, and chemical effects, mostly cavitation-related. This mechanism increases nutrient uptake, such as phosphorus and nitrogen, by facilitating the enhancement of algal cell membrane permeability, which leads to a more than 96% removal rate. Ultrasound generates oxidative radicals, for example, ·OH, that react with organic impurities and physically generate microjets and shock waves that enhance mass transfer and exposure of reaction sites. Ultrasound can be combined with other therapies to enhance the degradation of pollutants and reduce membrane fouling, including electrochemical processes, photocatalytic oxidation, and Fenton-type reactions. The prospects for improving performance and energy efficiency with advanced systems using dual or triple frequencies are promising. Ultrasound reduces secondary emissions and lowers chemical dependency, but its high-energy consumption prevents it from being used on a large scale in industry (Wang et al. 2023).

3.10 Energy Consumption Control for Chemical Transformations

The importance of efficient energy use in chemical transformations is critical to green and sustainable chemistry. It emphasizes the need for sustainable resource and energy use, with several of the fourteen articles in the special volume dedicated to this topic. These articles explore strategies for optimizing energy consumption in chemical processes, integrating renewable energy sources, and adopting innovative chemical engineering practices to reduce the energy footprint. The overarching goal is to transition toward more sustainable industrial processes that minimize environmental impact while maintaining efficiency (Lozano et al. 2016).

A great emphasis on energy intensity reduction in distillation processes has also been placed. New configurations, such as the FluxMax design, combine heat transfer and phase separation, so big energy is saved. Energy can be reduced to a minimum of 64% by enhancing exchange between the vapor and liquid phases, while additional heat exchanger surfaces might have to be added for this effect. Such energy-efficient column designs contribute toward achieving industry climate goals, primarily by reducing emitted CO_2 (Parekh et al. 2023).

In addition, high efficiency in terms of energy optimization in the chemical process industries calls for a techno-managerial approach. Energy-saving techniques can be integrated with a management approach for businesses to reduce operational costs and energy utilization. This kind of energy management system illustrates how energy-intensive industries, such as the chemical and petroleum industry sectors, can reduce their energy use by as much as 40%. This hybrid approach ensures long-term energy sustainability and efficiency (Majeed et al. 2023).

Various industry strategies and approaches have been invented to control energy usage during chemical reactions with a strong focus on energy intensity and process intensification. Improvement of existing manufacturing processes is an important method. For example, the Dow Chemical Company has decreased energy intensity by 22% since 1994. This is already considered a prime achievement. This was achieved through an organized program that improved system efficiency, reduced waste, and optimized operating conditions (He et al. 2023).

Optimization of duty cycles as well as ratios of material to solvent, for instance, presents opportunities in sonochemistry for the reduction of energy use. Adjusting the duty cycle appears to allow for a decrease in energy use during ultrasound-assisted extraction without impairing efficiency. For example, a 50% duty cycle was optimal for several processes, offering significantly high-energy savings without impairment and yields of good extraction. Lowering the duty cycle will reduce the opportunity for overheating and thus increase energy consumption, helping to prevent severe energy waste (Manickam et al. 2023).

3.11 Conclusion

Sonochemistry represents a unique and promising platform for the development of chemical engineering, as it utilizes acoustic cavitation to augment the rates of reaction, material production, and the productivity of associated processes. The dynamics of the bubble phase are relevant to the reactor design and operation, affecting various chemical and physical changes. The examples of the synthesis of nanoparticles, nanoemulsions, membrane processes, and water treatment show the versatility of the technique in various industries. In addition, the development of ultrasound generation and mass transfer models presents relevant information to improve the processes. It has become more essential and revealed a whole new facet of sonochemistry wherein the principles seamlessly merge with applications and make sonochemistry the flexible weapon of sustainable and efficient chemical engineering technologies.

References

Abbondanza D, Gallo M, Casciola CM (2023) Cavitation over solid surfaces: microbubble collapse, shock waves, and elastic response. Meccanica 58:1109–1119. https://doi.org/10.1007/s11012-022-01606-5

Adamou P, Harkou E, Villa A, Constantinou A, Dimitratos N (2024) Ultrasonic reactor set-ups and applications: a review. Ultrason Sonochem 107:106925. https://doi.org/10.1016/j.ultsonch.2024.106925

Afshari A, Ekberg L, Forejt L, Mo J, Rahimi S, Siegel J, Chen W, Wargocki P, Zurami S, Zhang J (2020) Electrostatic precipitators as an indoor air cleaner—a literature review. Sustainability 12:8774. https://doi.org/10.3390/su12218774

Agarwal A, Rai SK, Lin Y-C, Patnaik RK, Yeh JA (2020) Ammonia selectivity over acetone by viscosity modulation of silicone oil filter for diagnosing liver dysfunction. ECS J Solid State Sci Technol 9:115030. https://doi.org/10.1149/2162-8777/abc513

Arya SS, More PR, Ladole MR, Pegu K, Pandit AB (2023) Non-thermal, energy efficient hydrodynamic cavitation for food processing, process intensification and extraction of natural bioactives: a review. Ultrason Sonochem 98:106504. https://doi.org/10.1016/j.ultsonch.2023.106504

Asakura Y, Maebayashi M, Koda S (2005) Study on efficiency and characterization in a cylindrical sonochemical reactor. J Chem Eng Jpn 38:1008–1014. https://doi.org/10.1252/jcej.38.1008

References

Asgharzadehahmadi S, Abdul Raman AA, Parthasarathy R, Sajjadi B (2016) Sonochemical reactors: review on features, advantages and limitations. Renew Sustain Energy Rev 63:302–314. https://doi.org/10.1016/j.rser.2016.05.030

Ashokkumar M (2011) The characterization of acoustic cavitation bubbles—an overview. Ultrasonics Sonochem 18:864–872. https://doi.org/10.1016/j.ultsonch.2010.11.016

Aslam J, Ali R, Gupta N, Moeed K (2024) Indoor air pollution—a slow poison

Baek S, Kim KY, Kim G, Yun TS (2024) Linear and nonlinear ultrasound parameters attributed to anisotropy in granite. Sci Rep 14:26986. https://doi.org/10.1038/s41598-024-78367-6

Bampouli A, Goris Q, Van Olmen J, Solmaz S, Noorul Hussain M, Stefanidis GD, Van Gerven T (2023) Understanding the ultrasound field of high viscosity mixtures: experimental and numerical investigation of a lab scale batch reactor. Ultrason Sonochem 97:106444. https://doi.org/10.1016/j.ultsonch.2023.106444

Banakar VV, Sabnis SS, Gogate PR, Raha A, Saurabh, (2022) Ultrasound assisted continuous processing in microreactors with focus on crystallization and chemical synthesis: a critical review. Chem Eng Res des 182:273–289. https://doi.org/10.1016/j.cherd.2022.03.049

Bao J, Guo S, Fan D, Cheng J, Zhang Y, Pang X (2023) Sonoactivated nanomaterials: a potent armament for wastewater treatment. Ultrason Sonochem 99:106569. https://doi.org/10.1016/j.ultsonch.2023.106569

Bargole S, George S, Kumar Saharan V (2019) Improved rate of transesterification reaction in biodiesel synthesis using hydrodynamic cavitating devices of high throat perimeter to flow area ratios. Chem Eng Process—Process Intensification 139:1–13. https://doi.org/10.1016/j.cep.2019.03.012

Bhangu SK, Gupta S, Ashokkumar M (2017) Ultrasonic enhancement of lipase-catalysed transesterification for biodiesel synthesis. Ultrason Sonochem 34:305–309. https://doi.org/10.1016/j.ultsonch.2016.06.005

Boubendir L, Chikh S, Tadrist L (2020) On the surface tension role in bubble growth and detachment in a micro-tube. Int J Multiph Flow 124:103196. https://doi.org/10.1016/j.ijmultiphaseflow.2019.103196

Carlton JS (2019) Chapter 9—Cavitation. In: Carlton JS (ed) Marine propellers and propulsion, 4th ed. Butterworth-Heinemann, pp 217–260. https://doi.org/10.1016/B978-0-08-100366-4.00009-2

Caruso MM, Davis DA, Shen Q, Odom SA, Sottos NR, White SR, Moore JS (2009) Mechanically-induced chemical changes in polymeric materials. Chem Rev 109:5755–5798. https://doi.org/10.1021/cr9001353

Chakinala AG, Gogate PR, Burgess AE, Bremner DH (2008) Treatment of industrial wastewater effluents using hydrodynamic cavitation and the advanced Fenton process. Ultrason Sonochem 15:49–54. https://doi.org/10.1016/j.ultsonch.2007.01.003

Chang M, Lee Z, Park S, Park S (2020) Characteristics of flash boiling and its effects on spray behavior in gasoline direct injection injectors: a review. Fuel 271:117600. https://doi.org/10.1016/j.fuel.2020.117600

Chatel G (2018) How sonochemistry contributes to green chemistry? Ultrason Sonochem 40:117–122. https://doi.org/10.1016/j.ultsonch.2017.03.029

Chatterjee T, Rahaman SK, Alam SM (2024) Sonochemical synthesis of metal–organic frameworks. In: Synthesis of metal-organic frameworks via water-based routes. Elsevier, pp 121–142. https://doi.org/10.1016/B978-0-323-95939-1.00015-0

Chen H, Xi C, Xu H, Zhang X, Xiao Z, Xu S, Bai G (2024) Ultrasonic-driven degradation of organic pollutants using piezoelectric catalysts WS_2/Bi_2WO_6 heterojunction composites. Chemosphere 364:143008. https://doi.org/10.1016/j.chemosphere.2024.143008

Chowdhury P, Viraraghavan T (2009) Sonochemical degradation of chlorinated organic compounds, phenolic compounds and organic dyes—a review. Sci Total Environ 407:2474–2492. https://doi.org/10.1016/j.scitotenv.2008.12.031

Cristaldi DA, Yanar F, Mosayyebi A, García-Manrique P, Stulz E, Carugo D, Zhang X (2018) Easy-to-perform and cost-effective fabrication of continuous-flow reactors and their application for nanomaterials synthesis. In: New biotechnology, implementation of microreactor technology in biotechnology—IMTB 2017 47, pp 1–7. https://doi.org/10.1016/j.nbt.2018.02.002

Crudo D, Bosco V, Cavaglià G, Grillo G, Mantegna S, Cravotto G (2016) Biodiesel production process intensification using a rotor-stator type generator of hydrodynamic cavitation. Ultrason Sonochem 33:220–225. https://doi.org/10.1016/j.ultsonch.2016.05.001

Dehghani MH, Karri RR, Koduru JR, Manickam S, Tyagi I, Mubarak NM, Suhas (2023) Recent trends in the applications of sonochemical reactors as an advanced oxidation process for the remediation of microbial hazards associated with water and wastewater: a critical review. Ultrasonics Sonochem 94:106302. https://doi.org/10.1016/j.ultsonch.2023.106302

Devos C, Bampouli A, Brozzi E, Stefanidis GD, Dusselier M, Van Gerven T, Kuhn S (2025) Ultrasound mechanisms and their effect on solid synthesis and processing: a review. Chem Soc Rev. https://doi.org/10.1039/D4CS00148F

Dong Z, Delacour C, Mc Carogher K, Udepurkar AP, Kuhn S (2020) Continuous ultrasonic reactors: design, mechanism and application. Materials 13:344. https://doi.org/10.3390/ma13020344

Dular M, Griessler-Bulc T, Gutierrez-Aguirre I, Heath E, Kosjek T, Krivograd Klemenčič A, Oder M, Petkovšek M, Rački N, Ravnikar M, Šarc A, Širok B, Zupanc M, Žitnik M, Kompare B (2016) Use of hydrodynamic cavitation in (waste)water treatment. Ultrason Sonochem 29:577–588. https://doi.org/10.1016/j.ultsonch.2015.10.010

Dzah CS, Duan Y, Zhang H, Wen C, Zhang J, Chen G, Ma H (2020) The effects of ultrasound assisted extraction on yield, antioxidant, anticancer and antimicrobial activity of polyphenol extracts: a review. Food Biosci 35:100547. https://doi.org/10.1016/j.fbio.2020.100547

Enomoto N, Okitsu K (2015) Chapter 8—Application of ultrasound in inorganic synthesis. In: Grieser F, Choi P-K, Enomoto N, Harada H, Okitsu K, Yasui K (eds) Sonochemistry and the acoustic bubble. Elsevier, Amsterdam, pp 187–206. https://doi.org/10.1016/B978-0-12-801530-8.00008-6

Entezari MH, Pétrier C, Devidal P (2003) Sonochemical degradation of phenol in water: a comparison of classical equipment with a new cylindrical reactor. Ultrason Sonochem 10:103–108. https://doi.org/10.1016/S1350-4177(02)00136-0

Estivi L, Brandolini A, Condezo-Hoyos L, Hidalgo A (2022) Impact of low-frequency ultrasound technology on physical, chemical and technological properties of cereals and pseudocereals. Ultrason Sonochem 86:106044. https://doi.org/10.1016/j.ultsonch.2022.106044

Faïd F, Contamine F, Wilhelm AM, Delmas H (1998) Comparison of ultrasound effects in different reactors at 20 kHz. Ultrasonics Sonochem 5:119–124. https://doi.org/10.1016/S1350-4177(98)00009-1

Fathordoobady F, Sannikova N, Guo Y, Singh A, Kitts DD, Pratap-Singh A (2021) Comparing microfluidics and ultrasonication as formulation methods for developing hempseed oil nanoemulsions for oral delivery applications. Sci Rep 11:72. https://doi.org/10.1038/s41598-020-79161-w

Fattahi K, Boffito DC, Robert E (2024) Quantifying the chemical activity of cavitation bubbles in a cluster. Sci Rep 14:7978. https://doi.org/10.1038/s41598-024-56906-5

Feng D, Ferrasse J-H, Soric A, Boutin O (2019) Bubble characterization and gas–liquid interfacial area in two phase gas–liquid system in bubble column at low Reynolds number and high temperature and pressure. Chem Eng Res des 144:95–106. https://doi.org/10.1016/j.cherd.2019.02.001

Frank M (1952) Underwater transducer. US2613261A

Fuchs FJ (2015) 19—Ultrasonic cleaning and washing of surfaces. In: Gallego-Juárez JA, Graff KF (eds) Power ultrasonics. Woodhead Publishing, Oxford, pp 577–609. https://doi.org/10.1016/B978-1-78242-028-6.00019-3

Gaikwad SG, Pandit AB (2008) Ultrasound emulsification: effect of ultrasonic and physicochemical properties on dispersed phase volume and droplet size. Ultrason Sonochem 15:554–563. https://doi.org/10.1016/j.ultsonch.2007.06.011

Garcia-Vargas I, Louisnard O, Barthe L (2023) Extensive investigation of geometric effects in sonoreactors: analysis by luminol mapping and comparison with numerical predictions. Ultrason Sonochem 99:106542. https://doi.org/10.1016/j.ultsonch.2023.106542

Georgogianni KG, Katsoulidis AK, Pomonis PJ, Manos G, Kontominas MG (2009) Transesterification of rapeseed oil for the production of biodiesel using homogeneous and heterogeneous catalysis. Fuel Process Technol 90:1016–1022. https://doi.org/10.1016/j.fuproc.2009.03.002

Ghasemi N, Gbeddy G, Egodawatta P, Zare F, Goonetilleke A (2020) Removal of polycyclic aromatic hydrocarbons from wastewater using dual-mode ultrasound system. Water Environ J 34:425–434. https://doi.org/10.1111/wej.12540

Ghayal D, Pandit AB, Rathod VK (2013) Optimization of biodiesel production in a hydrodynamic cavitation reactor using used frying oil. Ultrason Sonochem 20:322–328. https://doi.org/10.1016/j.ultsonch.2012.07.009

Gilmanova L, Bon V, Shupletsov L, Pohl D, Rauche M, Brunner E, Kaskel S (2021) Chemically stable carbazole-based imine covalent organic frameworks with acidochromic response for humidity control applications. J Am Chem Soc 143:18368–18373. https://doi.org/10.1021/jacs.1c07148

Gogate PR, Patil PN (2016) Sonochemical reactors. Top Curr Chem (z) 374:61. https://doi.org/10.1007/s41061-016-0064-9

Gogate PR, Prajapat AL (2015) Depolymerization using sonochemical reactors: a critical review. Ultrason Sonochem 27:480–494. https://doi.org/10.1016/j.ultsonch.2015.06.019

Goswami KP, Pugazhenthi G (2020) Credibility of polymeric and ceramic membrane filtration in the removal of bacteria and virus from water: a review. J Environ Manage 268:110583. https://doi.org/10.1016/j.jenvman.2020.110583

Han S, Huang Y, Huang T, Li Y (2024) Study on bubble collapse mechanism in aeration system based on venturi cavitation effect. Process Saf Environ Prot 185:940–946. https://doi.org/10.1016/j.psep.2024.03.092

He L-L, Liu X-P, Wang Y-X, Wang Z-X, Yang Y-J, Gao Y-P, Liu B, Wang X (2016) Sonochemical degradation of methyl orange in the presence of Bi_2WO_6: effect of operating parameters and the generated reactive oxygen species. Ultrason Sonochem 33:90–98. https://doi.org/10.1016/j.ultsonch.2016.04.028

He X, He H, Barzagli F, Amer MW, Li C, Zhang R (2023) Analysis of the energy consumption in solvent regeneration processes using binary amine blends for CO_2 capture. Energy 270:126903. https://doi.org/10.1016/j.energy.2023.126903

Hofmann M, Haeberlin A, De Brot S, Stahel A, Keppner H, Burger J (2023) Development and evaluation of a titanium-based planar ultrasonic scalpel for precision surgery. Ultrasonics 130:106927. https://doi.org/10.1016/j.ultras.2023.106927

Hong WC, Kim YY, Kwon CD, So KC (2023) Effects of ultrasonic power and intensity of mechanical agitation on pretreatment of a gold-bearing arsenopyrite. Arch Acous:419–428. https://doi.org/10.24425/aoa.2024.148801

Hoo DY, Low ZL, Low DYS, Tang SY, Manickam S, Tan KW, Ban ZH (2022) Ultrasonic cavitation: an effective cleaner and greener intensification technology in the extraction and surface modification of nanocellulose. Ultrason Sonochem 90:106176. https://doi.org/10.1016/j.ultsonch.2022.106176

Hu Y-T, Ting Y, Hu J-Y, Hsieh S-C (2017) Techniques and methods to study functional characteristics of emulsion systems. J Food Drug Anal Diet Nat Compounds 25:16–26. https://doi.org/10.1016/j.jfda.2016.10.021

Husseini GA, Diaz de la Rosa MA, Richardson ES, Christensen DA, Pitt WG (2005) The role of cavitation in acoustically activated drug delivery. J Control Release 107:253–261. https://doi.org/10.1016/j.jconrel.2005.06.015

Janer M, Plantà X, Riera D (2020) Ultrasonic moulding: current state of the technology. Ultrasonics 102:106038. https://doi.org/10.1016/j.ultras.2019.106038

Jiang Q, Zhang M, Xu B (2020) Application of ultrasonic technology in postharvested fruits and vegetables storage: a review. Ultrason Sonochem 69:105261. https://doi.org/10.1016/j.ultsonch.2020.105261

Jothinathan L, Cai QQ, Ong SL, Hu JY (2021) Organics removal in high strength petrochemical wastewater with combined microbubble-catalytic ozonation process. Chemosphere 263:127980. https://doi.org/10.1016/j.chemosphere.2020.127980

Kaci M, Belhaffef A, Meziane S, Dostert G, Menu P, Velot É, Desobry S, Arab-Tehrany E (2018) Nanoemulsions and topical creams for the safe and effective delivery of lipophilic antioxidant coenzyme Q10. Colloids Surf, B 167:165–175. https://doi.org/10.1016/j.colsurfb.2018.04.010

Kalita U, Jafari VF, Ashokkumar M, Singha NK, Qiao GG (2024) Synthesis of ultra-high molecular weight homo- and copolymers via an ultrasonic emulsion process with a fast rate. Commun Chem 7:1–9. https://doi.org/10.1038/s42004-024-01191-6

Kamøy B, Magno M, Nøland ST, Moe MC, Petrovski G, Vehof J, Utheim TP (2022) Video display terminal use and dry eye: preventive measures and future perspectives. Acta Ophthalmol 100:723–739. https://doi.org/10.1111/aos.15105

Kentish S, Wooster TJ, Ashokkumar M, Balachandran S, Mawson R, Simons L (2008) The use of ultrasonics for nanoemulsion preparation. In: Innovative food science & emerging technologies, food innovation: emerging science, technologies and applications (FIESTA) conference 9, pp 170–175. https://doi.org/10.1016/j.ifset.2007.07.005

Khadhraoui B, Ummat V, Tiwari BK, Fabiano-Tixier AS, Chemat F (2021) Review of ultrasound combinations with hybrid and innovative techniques for extraction and processing of food and natural products. Ultrason Sonochem 76:105625. https://doi.org/10.1016/j.ultsonch.2021.105625

Khairiyah AN, Sugandi G, Kurniadi D (2023) Design and characterization of ultrasonic langevin transducer 20 kHz using a stepped horn front-mass. J Eng Technol Sci 55:357–372. https://doi.org/10.5614/j.eng.technol.sci.2023.55.4.1

Kojima Y, Takai S (2019) Transesterification of vegetable oil with methanol using solid base catalyst of calcium oxide under ultrasonication. Chem Eng Process Process Intensification 136:101–106. https://doi.org/10.1016/j.cep.2018.12.007

Kormušoski A, Lazarevska AM, Gečevska V (2024) Needs assessment of ambient CO_2 monitoring solution. MESJ 42:53–60. https://doi.org/10.55302/MESJ24421053k

Kumar A, Gogate PR, Pandit AB, Delmas H, Wilhelm AM (2004) Gas−liquid mass transfer studies in sonochemical reactors. Ind Eng Chem Res 43:1812–1819. https://doi.org/10.1021/ie0341146

Kumar K, Srivastav S, Sharanagat VS (2020) Ultrasound assisted extraction (UAE) of bioactive compounds from fruit and vegetable processing by-products: a review. Ultrason Sonochem 70:105325. https://doi.org/10.1016/j.ultsonch.2020.105325

Kwon G, Cho D-W, Park J, Bhatnagar A, Song H (2023) A review of plastic pollution and their treatment technology: a circular economy platform by thermochemical pathway. Chem Eng J 464:142771. https://doi.org/10.1016/j.cej.2023.142771

Laermer F, Franssila S, Sainiemi L, Kolari K (2020) Deep reactive ion etching. In: Handbook of silicon based MEMS materials and technologies. Elsevier, pp 417–446. https://doi.org/10.1016/B978-0-12-817786-0.00016-5

Lei Y, Hou J, Fang C, Tian Y, Naidu R, Zhang J, Zhang X, Zeng Z, Cheng Z, He J, Tian D, Deng S, Shen F (2023) Ultrasound-based advanced oxidation processes for landfill leachate treatment: energy consumption, influences, mechanisms and perspectives. Ecotoxicol Environ Saf 263:115366. https://doi.org/10.1016/j.ecoenv.2023.115366

Li F (2024) Lead-based piezoelectric materials. In: Wu J (ed) Piezoelectric materials. Wiley, pp 33–49. https://doi.org/10.1002/9783527841233.ch3

Li Y, Khan NF (2016) Ultrasonic lens cleaning system with current sensing. US20160266379A1

Li T, Siegel JA (2021) The impact of control strategies on filtration performance. Energy Buildings 252:111378. https://doi.org/10.1016/j.enbuild.2021.111378

Li N, Wang C, Jia N, Ma Z, Dang Y, Sun C, Du H, Xu Z, Li F (2024) A novel method to fabricate curved piezoelectric composites with high piezoelectric phase volume fraction. Ceram Int 50:38911–38916. https://doi.org/10.1016/j.ceramint.2024.07.255

Liu K, Jing B, Kang J, Han L, Chang J (2024a) Ultrasound-enabled nanomedicine for tumor theranostics. Engineering. https://doi.org/10.1016/j.eng.2024.01.030

Liu P, Zhang X, Liu G, Hao Lim S, Pun Wan M, Lisak G, Feng Ng B (2024b) Ultrasonic aerosol agglomeration: manipulation of particle deposition and its impact on air filter pressure drop. Ultrason Sonochem 103:106774. https://doi.org/10.1016/j.ultsonch.2024.106774

Lozano FJ, Freire P, Guillén-Gozalbez G, Jiménez-Gonzalez C, Sakao T, Dowell NM, Ortiz MG, Trianni A, Carpenter A, Viveros T (2016) New perspectives for sustainable resource and energy use, management and transformation: approaches from green and sustainable chemistry and engineering. J Clean Prod 118:1–3. https://doi.org/10.1016/j.jclepro.2016.01.041

Mahamuni NN, Adewuyi YG (2010) Advanced oxidation processes (AOPs) involving ultrasound for waste water treatment: a review with emphasis on cost estimation. Ultrason Sonochem Sonochem Scale up Industr Dev 17:990–1003. https://doi.org/10.1016/j.ultsonch.2009.09.005

Mahmood T, Akhtar N, Manickam S (2014) Interfacial film stabilized W/O/W nano multiple emulsions loaded with green tea and lotus extracts: systematic characterization of physicochemical properties and shelf-storage stability. J Nanobiotechnol 12:20. https://doi.org/10.1186/1477-3155-12-20

Mahvi A (2009) Application of ultrasonic technology for water and wastewater treatment. Iranian J Publ Health 38:1–17

Majeed Y, Khan MU, Waseem M, Zahid U, Mahmood F, Majeed F, Sultan M, Raza A (2023) Renewable energy as an alternative source for energy management in agriculture. Energy Rep 10:344–359. https://doi.org/10.1016/j.egyr.2023.06.032

Manickam S, Camilla Boffito D, Flores EMM, Leveque J-M, Pflieger R, Pollet BG, Ashokkumar M (2023) Ultrasonics and sonochemistry: editors' perspective. Ultrason Sonochem 99:106540. https://doi.org/10.1016/j.ultsonch.2023.106540

Marhasin E (2005) Ultrasonic reactor and process for ultrasonic treatment of materials. US20050260106A1

Mark G, Tauber A, Laupert R, Schuchmann H-P, Schulz D, Mues A, von Sonntag C (1998) OH-radical formation by ultrasound in aqueous solution—Part II: Terephthalate and Fricke dosimetry and the influence of various conditions on the sonolytic yield. Ultrason Sonochem 5:41–52. https://doi.org/10.1016/S1350-4177(98)00012-1

Marwa A, Iskandarsyah, Jufri M (2023) Nanoemulsion curcumin injection showed significant anti-inflammatory activities on carrageenan-induced paw edema in Sprague-Dawley rats. Heliyon 9:e15457. https://doi.org/10.1016/j.heliyon.2023.e15457

Masiello D, Tudela I, Shaw SJ, Jacobson B, Prentice P, Valluri P, Govindarajan R (2024) Mass and heat transfer in audible sound driven bubbles. Ultrason Sonochem 111:107068. https://doi.org/10.1016/j.ultsonch.2024.107068

Mason TJ (2003) Sonochemistry and sonoprocessing: the link, the trends and (probably) the future. Ultrasonics Sonochem Sel Papers Eighth Conf Eur Soc Sonochem 10:175–179. https://doi.org/10.1016/S1350-4177(03)00086-5

Masoumi HRF, Kassim A, Basri M, Abdullah DK (2011) Determining optimum conditions for lipase-catalyzed synthesis of triethanolamine (TEA)-based esterquat cationic surfactant by a Taguchi Robust design method. Molecules 16:4672–4680. https://doi.org/10.3390/molecules16064672

Merouani S, Dehane A, Hamdaoui O (2024) Ultrasonic destruction of surfactants. Ultrason Sonochem 109:107009. https://doi.org/10.1016/j.ultsonch.2024.107009

Mi S, Zhu C, Ma Y, Fu T (2022) Bubble formation in high-viscosity liquids in step-emulsification microdevices. J Ind Eng Chem 114:221–232. https://doi.org/10.1016/j.jiec.2022.07.012

Modarres-Gheisari SMM, Gavagsaz-Ghoachani R, Malaki M, Safarpour P, Zandi M (2019) Ultrasonic nano-emulsification—a review. Ultrason Sonochem 52:88–105. https://doi.org/10.1016/j.ultsonch.2018.11.005

Mohammadi AR, Mehrdadi N, Bidhendi GN, Torabian A (2011) Excess sludge reduction using ultrasonic waves in biological wastewater treatment. Desalination 275:67–73. https://doi.org/10.1016/j.desal.2011.02.030

Mohod AV, Gogate PR, Viel G, Firmino P, Giudici R (2017) Intensification of biodiesel production using hydrodynamic cavitation based on high speed homogenizer. Chem Eng J 316:751–757. https://doi.org/10.1016/j.cej.2017.02.011

Moholkar VS (2015) Mathematical models for sonochemical effects induced by hydrodynamic cavitation. In: Ashokkumar M (ed) Handbook of ultrasonics and sonochemistry. Springer, Singapore, pp 1–48. https://doi.org/10.1007/978-981-287-470-2_51-1

Moholkar VS, Senthil Kumar P, Pandit AB (1999) Hydrodynamic cavitation for sonochemical effects. Ultrason Sonochem 6:53–65. https://doi.org/10.1016/S1350-4177(98)00030-3

Nabi BG, Mukhtar K, Ansar S, Hassan SA, Hafeez MA, Bhat ZF, Mousavi Khaneghah A, Haq AU, Aadil RM (2024) Application of ultrasound technology for the effective management of waste from fruit and vegetable. Ultrason Sonochem 102:106744. https://doi.org/10.1016/j.ultsonch.2023.106744

Obaideen K, Shehata N, Sayed ET, Abdelkareem MA, Mahmoud MS, Olabi AG (2022) The role of wastewater treatment in achieving sustainable development goals (SDGs) and sustainability guideline. Energy Nexus 7:100112. https://doi.org/10.1016/j.nexus.2022.100112

Ohl S-W, Klaseboer E, Khoo BC (2015) Bubbles with shock waves and ultrasound: a review. Interface Focus 5:20150019. https://doi.org/10.1098/rsfs.2015.0019

Okoli CU, Kuttiyiel KA, Cole J, McCutchen J, Tawfik H, Adzic RR, Mahajan D (2018) Solvent effect in sonochemical synthesis of metal-alloy nanoparticles for use as electrocatalysts. Ultrason Sonochem 41:427–434. https://doi.org/10.1016/j.ultsonch.2017.09.049

Olaya-Escobar D-R, Quintana-Jiménez L-A, González-Jiménez E-E, Olaya-Escobar E-S (2020) Ultrasound applied in the reduction of viscosity of heavy crude oil. Facultad de Ingeniería 29

Omelchenko E, Trushkova E, Sitnik S, Bogatina A (2022) Study of the effectiveness of innovative air purification systems used in the design of road construction enterprises. Transp Res Procedia 61:594–599. https://doi.org/10.1016/j.trpro.2022.01.096

Orthaber U, Zevnik J, Petkovšek R, Dular M (2020) Cavitation bubble collapse in a vicinity of a liquid-liquid interface—basic research into emulsification process. Ultrason Sonochem 68:105224. https://doi.org/10.1016/j.ultsonch.2020.105224

Parekh N, Kurian J, Patil R, Gautam R (2023) A review on techno managerial approaches to energy optimization in chemical process industries. Front Energy Res 10:1107912. https://doi.org/10.3389/fenrg.2022.1107912

Peng K, Qin FGF, Jiang R, Kang S (2020) Interpreting the influence of liquid temperature on cavitation collapse intensity through bubble dynamic analysis. Ultrason Sonochem 69:105253. https://doi.org/10.1016/j.ultsonch.2020.105253

Pereira B, Arantes V (2018) Chapter 9—Nanocelluloses from sugarcane biomass. In: Chandel AK, Luciano Silveira MH (eds) Advances in sugarcane biorefinery. Elsevier, pp 179–196. https://doi.org/10.1016/B978-0-12-804534-3.00009-4

Peterson ME, Daniel RM, Danson MJ, Eisenthal R (2007) The dependence of enzyme activity on temperature: determination and validation of parameters. Biochem J 402:331–337. https://doi.org/10.1042/BJ20061143

Pham TD, Shrestha RA, Virkutyte J, Sillanpää M (2013) Recent studies in environmental applications of ultrasound. J Environ Eng Sci 8:403–412. https://doi.org/10.1680/jees.2013.0040

Pokhrel N, Vabbina PK, Pala N (2016) Sonochemistry: science and engineering. Ultrason Sonochem 29:104–128. https://doi.org/10.1016/j.ultsonch.2015.07.023

Preeti, Sambhakar S, Malik R, Bhatia S, Al Harrasi A, Rani C, Saharan R, Kumar S, Geeta, Sehrawat R (2023) Nanoemulsion: an emerging novel technology for improving the bioavailability of drugs. Scientifica (Cairo) 2023:6640103. https://doi.org/10.1155/2023/6640103

Pukale DD, Maddikeri GL, Gogate PR, Pandit AB, Pratap AP (2015) Ultrasound assisted transesterification of waste cooking oil using heterogeneous solid catalyst. Ultrason Sonochem 22:278–286. https://doi.org/10.1016/j.ultsonch.2014.05.020

Qasim M, Darwish NN, Mhiyo S, Darwish NA, Hilal N (2018) The use of ultrasound to mitigate membrane fouling in desalination and water treatment. Desalination 443:143–164. https://doi.org/10.1016/j.desal.2018.04.007

Ramezani M, Kong B, Gao X, Olsen MG, Vigil RD (2015) Experimental measurement of oxygen mass transfer and bubble size distribution in an air–water multiphase Taylor-Couette vortex bioreactor. Chem Eng J 279:286–296. https://doi.org/10.1016/j.cej.2015.05.007

Rennoll V, McLane I, Eisape A, Grant D, Hahn H, Elhilali M, West JE (2023) Electrostatic acoustic sensor with an impedance-matched diaphragm characterized for body sound monitoring. ACS Appl Bio Mater 6:3241–3256. https://doi.org/10.1021/acsabm.3c00359

Richa D, Kumar R, Shukla R, Khan K (2020) Ultrasound assisted essential oil extraction technology: new boon in food industry. 22:81–85

Saif Ur Rehman M, Kim I, Chisti Y, Han J-I (2013) Use of ultrasound in the production of bioethanol from lignocellulosic biomass. Energy Educ Sci Technol 30:1391–1410

Sajjadi B, Asgharzadehahmadi S, Asaithambi P, Raman AAA, Parthasarathy R (2017) Investigation of mass transfer intensification under power ultrasound irradiation using 3D computational simulation: a comparative analysis. Ultrason Sonochem 34:504–518. https://doi.org/10.1016/j.ultsonch.2016.06.026

Sakthipandi K, Sethuraman B, Venkatesan K, Alhashmi B, Purushothaman G, Ansari IA (2024) Ultrasound-based sonochemical synthesis of nanomaterials. In: Garg N, Gautam C, Rab S, Wan M, Agarwal R, Yadav S (eds) Handbook of vibroacoustics, noise and harshness. Springer Nature, Singapore, pp 1–46. https://doi.org/10.1007/978-981-99-4638-9_58-1

Saleh TA (2021) Chapter 2—Materials: types and general classifications. In: Saleh TA (ed) Polymer hybrid materials and nanocomposites, plastics design library. William Andrew Publishing, pp 27–58. https://doi.org/10.1016/B978-0-12-813294-4.00008-X

Sancheti SV, Gogate PR (2017) A review of engineering aspects of intensification of chemical synthesis using ultrasound. Ultrason Sonochem 36:527–543. https://doi.org/10.1016/j.ultsonch.2016.08.009

Sandhya M, Ramasamy D, Sudhakar K, Kadirgama K, Harun WSW (2021) Ultrasonication an intensifying tool for preparation of stable nanofluids and study the time influence on distinct properties of graphene nanofluids—a systematic overview. Ultrason Sonochem 73:105479. https://doi.org/10.1016/j.ultsonch.2021.105479

Santha Kumar ARS, Padmakumar A, Kalita U, Samanta S, Baral A, Singha NK, Ashokkumar M, Qiao GG (2023) Ultrasonics in polymer science: applications and challenges. Prog Mater Sci 136:101113. https://doi.org/10.1016/j.pmatsci.2023.101113

Scardina P (n.d.) Effects of dissolved gas supersaturation and bubble formation on water treatment plant performance

Seike Y, Takagi R, Hikita T, Honda Y, Taoka N, Ichino Y, Mori T (2024) Evaluating the removal of polystyrene latex particles using the quartz transducer ultrasonic cleaners in the megahertz frequency range. ECS Trans 114:83. https://doi.org/10.1149/11401.0083ecst

Sharifishourabi M, Dincer I, Mohany A (2024) Implementation of experimental techniques in ultrasound-driven hydrogen production: a comprehensive review. Int J Hydrogen Energy 62:1183–1204. https://doi.org/10.1016/j.ijhydene.2024.03.013

Sharma SK, Saxena KK, Salem KH, Mohammed KA, Singh R, Prakash C (2024) Effects of various fabrication techniques on the mechanical characteristics of metal matrix composites: a review. Adv Mater Process Technol 10:277–294. https://doi.org/10.1080/2374068X.2022.2144276

Shen L, Pang S, Zhong M, Sun Y, Qayum A, Liu Y, Rashid A, Xu B, Liang Q, Ma H, Ren X (2023a) A comprehensive review of ultrasonic assisted extraction (UAE) for bioactive components: principles, advantages, equipment, and combined technologies. Ultrason Sonochem 101:106646. https://doi.org/10.1016/j.ultsonch.2023.106646

Shen Y, Pflieger R, Chen W, Ashokkumar M (2023b) The effect of bulk viscosity on single bubble dynamics and sonoluminescence. Ultrason Sonochem 93:106307. https://doi.org/10.1016/j.ultsonch.2023.106307

Shi Y, Shi ZM (2021) Surface treatment of cementitious composites by ultrasound and its effect on durability performance. J Mater Civ Eng 33:04020487. https://doi.org/10.1061/(ASCE)MT.1943-5533.0003575

Sims CC (1960) Bubble transducer for radiating high-power low frequency sound in water. J Acous Soc Am 32:1305–1308. https://doi.org/10.1121/1.1907899

Sivakumar M, Tang SY, Tan KW (2014) Cavitation technology—a greener processing technique for the generation of pharmaceutical nanoemulsions. Ultrasonics Sonochem AOSS 2013(21):2069–2083. https://doi.org/10.1016/j.ultsonch.2014.03.025

Somanathan A, Mathew N, Arfin T (2024) Environmental impacts and developments in waste-derived nanoparticles for air pollution control. In: Waste-derived nanoparticles. Elsevier, pp 281–318. https://doi.org/10.1016/B978-0-443-22337-2.00018-X

Song D, Seibert AF, Rochelle GT (2017) Effect of liquid viscosity on mass transfer area and liquid film mass transfer coefficient for GT-OPTIMPAK 250Y. Energy Procedia 114:2713–2727. https://doi.org/10.1016/j.egypro.2017.03.1534

Speight JG (2019) 9—Gas condensate. In: Speight JG (ed) Natural gas, 2nd ed. Gulf Professional Publishing, Boston, pp 325–358. https://doi.org/10.1016/B978-0-12-809570-6.00009-6

Sreeharsha N, Kunigal Sridhar S, Bhuvanahalli Rangappa A, Goudanavar P, Karadigere Nagaraju P, Raghavendra Naveen N, Narayanappa Shiroorkar P, Haq Asif A, Meravanige G, Swaroop Duddi Sreehari K (2024) Ultrasonication-mediated synthesis of diblock polymer-based nanoparticles for advanced drug delivery systems: insights and optimization. Ultrason Sonochem 111:107137. https://doi.org/10.1016/j.ultsonch.2024.107137

Storey BD, Szeri AJ (2001) A reduced model of cavitation physics for use in sonochemistry. Proc R Soc Lond. Ser A: Math Phys Eng Sci 457:1685–1700. https://doi.org/10.1098/rspa.2001.0784

Sullivan JF (2009) Ultrasonic cleaner. EP2026913A2

Suslick KS (2003) 1.41—Sonochemistry. In: McCleverty JA, Meyer TJ (eds) Comprehensive coordination chemistry II. Pergamon, Oxford, pp 731–739. https://doi.org/10.1016/B0-08-043748-6/01046-X

Tang SY, Sivakumar M, Ng AM-H, Shridharan P (2012) Anti-inflammatory and analgesic activity of novel oral aspirin-loaded nanoemulsion and nano multiple emulsion formulations generated using ultrasound cavitation. Int J Pharm 430:299–306. https://doi.org/10.1016/j.ijpharm.2012.03.055

Tanuku S, Velisila D, Thatraju D, Vadaga AK (2024) Nanoemulsion formulation strategies for enhanced drug delivery: review article. J Pharma Insights Res 2:125–138. https://doi.org/10.69613/3f8m9151

Theerthagiri J, Madhavan J, Lee SJ, Choi MY, Ashokkumar M, Pollet BG (2020) Sonoelectrochemistry for energy and environmental applications. Ultrason Sonochem 63:104960. https://doi.org/10.1016/j.ultsonch.2020.104960

Torresin S, Aletta F, Oberman T, Vinciotti V, Albatici R, Kang J (2023) Measuring, representing and analysing indoor soundscapes: a data collection campaign in residential buildings with natural and mechanical ventilation in England. Build Environ 243:110726. https://doi.org/10.1016/j.buildenv.2023.110726

Torrisi L, Silipigni L, Torrisi A, Cutroneo M (2024) Luminescence in laser-generated functionalized carbon dots. Opt Laser Technol 177:111089. https://doi.org/10.1016/j.optlastec.2024.111089

Upendran S, Jose A (2022) Development of a portable ultraviolet germicidal irradiation system for controlling air borne infection. Int J Eng Res Technol (IJERT) 11:96–101

Verdini F, Crudo D, Bosco V, Kamler AV, Cravotto G, Calcio Gaudino E (2024) Advanced processes in water treatment: synergistic effects of hydrodynamic cavitation and cold plasma on Rhodamine B dye degradation. Processes 12:2128. https://doi.org/10.3390/pr12102128

Vernès L, Vian M, Chemat F (2020) Chapter 12—Ultrasound and microwave as green tools for solid-liquid extraction. In: Poole CF (ed) Liquid-phase extraction, handbooks in separation science. Elsevier, pp 355–374. https://doi.org/10.1016/B978-0-12-816911-7.00012-8

Wang H, Lu Y, Zhu J (2003) Preparation of cube-shaped CdS nanoparticles by sonochemical method. In: AsiaNano 2002. World Scientific, pp 63–67. https://doi.org/10.1142/9789812796714_0011

Wang D, Yan L, Ma X, Wang W, Zou M, Zhong J, Ding T, Ye X, Liu D (2018) Ultrasound promotes enzymatic reactions by acting on different targets: enzymes, substrates and enzymatic reaction systems. Int J Biol Macromol 119:453–461. https://doi.org/10.1016/j.ijbiomac.2018.07.133

Wang Y, Peng H, He X, Zhang J (2022) Cavitation bubbles with a tunable-surface-tension thermal lattice Boltzmann model. Phys Fluids 34:102008. https://doi.org/10.1063/5.0113500

Wang N, Li L, Wang K, Huang X, Han Y, Ma X, Wang M, Lv X, Bai X (2023) Study and application status of ultrasound in organic wastewater treatment. Sustainability 15:15524. https://doi.org/10.3390/su152115524

Wen H, Cheng D, Chen Y, Yue W, Zhang Z (2024) Review on ultrasonic technology enhanced biological treatment of wastewater. Sci Total Environ 925:171260. https://doi.org/10.1016/j.scitotenv.2024.171260

Wood RJ, Lee J, Bussemaker MJ (2017) A parametric review of sonochemistry: control and augmentation of sonochemical activity in aqueous solutions. Ultrason Sonochem 38:351–370. https://doi.org/10.1016/j.ultsonch.2017.03.030

Wu TY, Guo N, Teh CY, Hay JXW (2013) Applications of ultrasound technology in environmental remediation. In: Wu TY, Guo N, Teh CY, Hay JXW (eds) Advances in ultrasound technology for environmental remediation. Springer Netherlands, Dordrecht, pp 13–93. https://doi.org/10.1007/978-94-007-5533-8_3

Wu WH, Eskin DG, Priyadarshi A, Subroto T, Tzanakis I, Zhai W (2021) New insights into the mechanisms of ultrasonic emulsification in the oil–water system and the role of gas bubbles. Ultrason Sonochem 73:105501. https://doi.org/10.1016/j.ultsonch.2021.105501

Xiang L, Fu M, Wang T, Wang D, Xv H, Miao W, Le T, Zhang L, Hu J (2024) Application and development of ultrasound in industrial crystallization. Ultrason Sonochem 111:107062. https://doi.org/10.1016/j.ultsonch.2024.107062

Xie L, Lian Y, Du F, Wang Y, Lu Z (2024) Optical methods of laser ultrasonic testing technology in the industrial and engineering applications: a review. Opt Laser Technol 176:110876. https://doi.org/10.1016/j.optlastec.2024.110876

Yamashita T, Ando K (2019) Low-intensity ultrasound induced cavitation and streaming in oxygen-supersaturated water: role of cavitation bubbles as physical cleaning agents. Ultrason Sonochem 52:268–279. https://doi.org/10.1016/j.ultsonch.2018.11.025

Yin J, Zhang Y, Zhu J, Zhang Y, Li S (2021) On the thermodynamic behaviors and interactions between bubble pairs: a numerical approach. Ultrason Sonochem 70:105297. https://doi.org/10.1016/j.ultsonch.2020.105297

Yuming L, Guoqing C, Xin F, Yanpeng W (2022) Review on the adsorption of airborne molecular contaminants in electronic industry cleanrooms, pp 1095–1103

Zaib Q, Jouiad M, Ahmad F (2019) Ultrasonic synthesis of carbon nanotube-titanium dioxide composites: process optimization via response surface methodology. ACS Omega 4:535–545. https://doi.org/10.1021/acsomega.8b02706

Zhang Q-Q, Jin R-C (2015) The application of low-intensity ultrasound irradiation in biological wastewater treatment: a review. Crit Rev Environ Sci Technol 45:2728–2761. https://doi.org/10.1080/10643389.2015.1046772

Zhang P, Lin R, Huang Z, Wang X, You H, Gao J, Wang X (2015) The research on the enhanced effect of dual-frequency. Presented at the first international conference on information sciences, machinery, materials and energy. Atlantis Press, pp 1872–1876. https://doi.org/10.2991/icismme-15.2015.384

Zhang X-M, Li F, Wang C-H, Hu J, Mo R-Y, Shen Z-Z, Guo J-Z, Lin S-Y (2023a) Bubble nucleation in spherical liquid cavity wrapped by elastic medium. Chinese Phys B 32:064303. https://doi.org/10.1088/1674-1056/acaa30

Zhang Y, Yuan L, Liu S, Zhang J, Yang M, Song Y (2023b) Molecular dynamics simulation of bubble nucleation and growth during CO_2 Huff-n-Puff process in a CO_2-heavy oil system. Geoenergy Sci Eng 227:211852. https://doi.org/10.1016/j.geoen.2023.211852

Zhang K, Gao G, Wang Y, Wang Y, Li J, Xiang D, Zhao B (2024) Integrated high-frequency piezoelectric transducer within radial critical dimension of piezoceramics. Int J Mech Sci 269:109070. https://doi.org/10.1016/j.ijmecsci.2024.109070

Zhao H, Mo H, Mao P, Ran R, Zhou W, Liao K (2024) Tape-casting fabrication techniques for garnet-based membranes in solid-state lithium-metal batteries: a comprehensive review. ACS Appl Mater Interfaces. https://doi.org/10.1021/acsami.4c18516

Zheng H, Zheng Y, Zhu J (2022) Recent developments in hydrodynamic cavitation reactors: cavitation mechanism, reactor design, and applications. Engineering 19:180–198. https://doi.org/10.1016/j.eng.2022.04.027

Zhou Y, Liu Y, Zhang M, Feng Z, Yu D-G, Wang K (2022) Electrospun nanofiber membranes for air filtration: a review. Nanomaterials 12:1077. https://doi.org/10.3390/nano12071077

Zhu N, Wei Z, Chen C, Xiong X, Xiong Y, Zeng Z, Wang W, Jiang J, Fan Y, Su C (2022) High water adsorption MOFs with optimized pore-nanospaces for autonomous indoor humidity control and pollutants removal. Angew Chem Int Ed 61:e202112097. https://doi.org/10.1002/anie.202112097

Zupanc M, Pandur Ž, Stepišnik Perdih T, Stopar D, Petkovšek M, Dular M (2019) Effects of cavitation on different microorganisms: the current understanding of the mechanisms taking place behind the phenomenon. A review and proposals for further research. Ultrason Sonochem 57:147–165. https://doi.org/10.1016/j.ultsonch.2019.05.009

Applications in Environmental Engineering 4

Abstract

Environmental pollution is concerning for both biotic and abiotic components on Earth. Due to the rapid growth of industries, lifestyle changes in humans, migration toward cities and improper waste disposal methods, the Earth poses a significant threat. There is a crucial need to address these issues and develop technologies for proper waste management and reduce environmental pollution. Sonochemistry is one of the green approaches and emerging fields that can effectively decontaminate the environment. Due to its versatility, the field finds its application in various areas of wastewater treatment, water purification, soil and sludge treatment. The chapter discusses the use of ultrasound in the above applications.

4.1 Introduction

Sonochemistry and its applications in environmental engineering are continuously evolving, offering new opportunities for improvement and innovation in the field. The vast applications of this technology are limited in the minds of researchers working in the domain. Several aspects of this field deal with the degradation of organic materials, thereby preventing pollution (Adewuyi 2001). The field is a boon for the environment as it deals with less hazardous compounds, has specific reactions with minimal toxic compounds generated, and requires less energy than the usual available processes and recycling of resources (Savun-Hekimoğlu 2020). As discussed earlier, sonochemistry involves the processes of acoustic cavitation, formation of hydroxy radicals, and implosion of bubbles. Due to this, toxic components that emerge from household waste, industries, municipal wastewater treatment plants, personal hygiene products, dyes, medicines, petrol, and mining are decomposed by ultrasound (Savun-Hekimoğlu 2020).

The chemical industries, textiles, pharmaceuticals, and construction sites are major contributors to environmental pollution. According to a recent study, in India, around 7–8 lakh people die due to an increase in temperature caused by environmental pollution, where more than 50% of air pollution is caused by air pollution (Sharma et al. 2022). The pharmaceutical industry contributes significantly to wastewater disposal in sewage (Halling-Sørensen et al. 1998). Several factors determine the fate of pharmaceutical anti-infective agents that enter the environment. Pharmaceuticals are reported to enter aquatic environments by discharge from treatment plants or direct disposal (untreated form) (Andreozzi et al. 2003). Disposal methods for pharmaceuticals depend on factors such as population, the number of hospitals, industrial activity, the standard of living, and local conditions. Drugs disposed of in hospitals and factories and household wastes in urban areas end up in sewage, carried to wastewater treatment plants via sewers. They combine with other contaminants and waste waters, eventually reaching aquatic life (Halling-Sørensen et al. 1998). Drugs consumed are eliminated in sewage or wastewater treatment plants along with human excreta, and they are found in unaltered form or as drug metabolites. If discarded by households, the used and expired medications end up in sewage and eventually in aquatic systems. Several medicines frequently leak into water surroundings (Halling-Sørensen et al. 1998). Surface water is found to be in contact with agricultural land, and hence, drugs/drug metabolites enter humans and animals via the food chain (Kummerer 2003).

Drugs enter surface water through runoff, drainage, and percolation into groundwater through the agricultural route. Animal excrement has been reported to contain anti-infective agents. Compounds are carried by manure application in landfills either as suspended particles or in the aqueous phase. Anti-infectives are released into agricultural soils over time, acting as environmental reservoirs before being released into waste water (Naddeo et al. 2012; Segura et al. 2009).

Hospital wastewater (HWW) is also contaminated with numerous environmental contaminants. The common pollutants that threaten the environment and biotic life include steroids, personal care products, micropollutants, e.g., phenols, polyphenols, etc. It is observed that pretreatment has been given to the WW from dental clinics and medical laboratories. The International Commission on Radiological Protection (ICRP) has developed strategies for treating radiation-emitting compounds found in HWW (Khan et al. 2021).

Like contaminants in water, air contamination also negatively affects the environment. The common pollutants found in the environment include polyaromatic hydrocarbons (PAH), harmful chemicals like formaldehyde, carbon monoxide, and dioxide (Wei et al. 2021; Hoang et al. 2023). Microplastics are the emerging class of pollutants that are formed by fibers and synthetic textile materials. They can potentially damage the respiratory system of animals and humans, apart from forming dust hazards. (Akanyange et al. 2021) There are also traces of biological contaminants in the air, including various microbacterial and pathogenic strains. They include the species of *Mycobacterium tuberculosis, Aspergillus fungi*, and influenza viruses. When affected by an individual, they become a source of infection and thus potentially affect environment at a later stage (Kumar et al. 2021).

4.2 Environmental Sonochemistry

Environmental contamination results from releasing unmineralized pollutants into water resources, such as effluents from the textile, pulp & paper, and pharmaceutical sectors. In addition, non-organic materials, such as agricultural pesticides overuse, pollute land and water responsible for the same. Negative impacts are caused by the discharge of these unneutralized pollutants into the environment, and introducing these chemicals into aquatic life affects the water bodies and risks human health. For instance, fish are feminized by the presence of 4-nonylphenol (an endocrine disruptor) in sewage in wastewater, which has become a major ecological conservation issue (Sathishkumar and Viswanathan 2016).

Industries produce significant amounts of hazardous waste due to industrial development, lifestyle changes, and the growing human population demands. The increase in waste generation significantly impacts human health, making it essential to treat or dispose of it properly. Conventional methods to depose this waste include landfilling, carbon adsorption, incineration, chemical treatment, etc. However, these alternatives are unsuitable in the long run as they harm Earth's environment, plants, and animal life. Hence, other alternatives, such as chemical-based oxidation processes, also known as advanced oxidation processes (APOs), are a recent boon. These processes produce hydroxyl radicals in large quantities for water treatment. As discussed earlier, the products formed are carbon dioxide, small chain compounds, and inorganic ions that are harmless to the biotic life and environment (Crini and Lichtfouse 2019).

Wastewater treatment (WWT) is crucial because untreated industrial wastewater can enter water bodies directly or through leaching. This water, which is used for drinking or agricultural purposes, may cause a variety of health problems and also hamper the economy and progress of humankind. Thus, there is a need for safe and effective WWT (Salgot and Folch 2018). Sonochemistry can be viewed as a green science that paves the way for sustainable wastewater treatment. A recent field in AOPs that is in use for the environment is cavitation, which involves using ultrasound (US) for wastewater treatment. Its application is in three categories: (a) using cavitation as the only WWT source, (b) combining cavitation with additional therapies (e.g., with AOP), and (c) reducing the amount of chemicals used in chemical treatment through cavitation (Adewu-Yi 2005).

The treatment of sewage sludge is one of the significant challenges for officials in urban cities. Anaerobic digestion is effective but has flaws, such as methane gas production and longer fermentation time (more than 15 days). The use of US at a frequency of 31 kHz at higher intensity was demonstrated to treat the sewage sludge in about a week and produced twice the amount of biogas produced by a single anaerobic cycle (Mason 1999). In industry, most suspended matter is made of smoke, mist, dust, and suspended particles of inorganic matter. This matter causes respiratory issues and creates pollution in the environment. Acoustic cavitation is essential in treating the waste generated and remediating the environment. The main principle is that nodes are formed when an acoustic/sonic wave is generated. Particles tend to go toward nodes and agglomerate there. As their size increases, they no longer travel between the nodes and fall to the ground. However, these devices consume much energy and are unsuitable for extremely fine particles. Hence, a modification of this device by fitting a radiating plate that generates vibrations in the plane is a good alternative for fine particles and foams. Using a high-power US has been demonstrated to give better efficiency (Mason 2007).

The above-mentioned industries generate large amounts of organic wastes and chemical substances, and their suitable disposal becomes an issue. The use of sonochemical reactions to degrade industrially generated phenols, benzenes, and nitrogen-containing heterocyclic compounds has been reported. The methods are validated using COD, BOD, and analytical instruments such as GC and GC–MS (Colarusso and Serpone 1996). The use of cavitation or AOP as a process individually is not sufficient. A combined approach involving various processes like Fenton reactions, ozonolysis, and sonophotocatalysis gives more desirable results (Sathishkumar and Viswanathan 2016).

As discussed earlier, sonochemistry principles for water treatment are widely used. However, there are certain specifications when it comes to the effective removal of pollutants from water. These include the pH of the solution, its temperature, frequency of US, presence of dissolved gases, etc. (Hu et al. 2025; Akiya and Savage 2002; Davies et al. 2014; Al-Rubaiey 2024).

4.3 Elimination of Hazardous Substances

Most of the chemicals that are found in the environment, especially in WW, are aromatic compounds containing phenols, benzene derivatives, pesticides and herbicides containing chlorinated compounds, azo derivatives, etc. Table 4.1 represents the sonochemical conditions required for the degradation of these aromatic reagents (Ince et al. 2001).

In a recent article published by Preeti Gupta et al. a comparative approach between different sonochemical combination methods was discussed for the degradation of polycyclic aromatic hydrocarbons that are commonly known as PAHs. They are a common source of pollutants that are generated by vegetative waste as well as by large industries. They include fused benzene ring systems such as phenanthrene, naphthalene, chrysene, anthracene, etc. They observed the trend that combinative sonochemical treatment (sonicator + photo-Fenton) gave good results that alone sonicator or sonicator combined with Fenton or photon source (Gupta et al. 2021).

4.3 Elimination of Hazardous Substances 95

Table 4.1 Different sonochemical methods used for the degradation of environmental pollutants

Sr. No.	Pollutant	Sonochemical method	Conditions for degradation	References
1	Phenol	Ultrasonic horn with applied fluid flow	Flow rate—24 mL/min Power—10 W	Wood et al. (2021)
2	Phenol	High-frequency US	Frequency—45 kHz; Power—200 W, Current—1.2 A, pH—5	Zhang et al. (2020)
3	Phenol	Microwave-assisted-Fenton reagent treatment	Microwave power density—0.97 W/mL, Fe^{2+} addition—0.2 mmol/L, H_2O_2 addition—16 mmol/L, pH—5 radiation time—7.11 min	Zhang et al. (2020)
4	Benzene sulphonic acid (10^{-4} M)	Sonolysis	Power—80 W, Frequency—350 kHz, pH—4.8	Thomas et al. (2020)
5	Para nitro phenol	US assisted by iron NPs produced by Jatropha leaf extract	Frequency—40 kHz, time—240 min	Rawat et al. (2021)
6	Benzene (9.35×10^{-3} M)	Hydrodynamic cavitation	Cavitation—For an initial 30 min, Pressure—2.4 bar	Thanekar et al. (2021)
7	Alkyl benzene sulphonate	UV/US	Frequency—42 kHz, Power—160 W, UV lamp power—150W, Wavelength—254 nm	Razavi et al. (2020)
8	BTEXs (benzene, toluene, ethylbenzene and xylenes)	US/O_3 with dual frequency US	P—40 to 120 kHz with O_3, Time—40 min	Fedorov et al. (2024)
9	BTEXs (benzene, toluene, ethylbenzene and xylenes)	US/O_3	P—80 to 200 kHz with O_3, Time—80 min	Fedorov et al. (2024)

(continued)

Table 4.1 (continued)

Sr. No.	Pollutant	Sonochemical method	Conditions for degradation	References
10	HCHs (petroleum hydrocarbons, carbamazepine, phenanthrene, Polychlorinated biphenyls, ΣHCHs = 404 mg kg^{-1})	US/Persulfate	Frequency—20 kHz; Power—0 to 245 W, Power density—0–91 W/L, T—22 ± 2 °C; pH—alkaline (NaOH) conditions; reaction Time—1 to 3 h; Reaction medium—Soil	Checa-Fernández et al. (2022)
11	Diclofenac	Sono-ozonation	Power—400 W; Frequency—20 kHz; Temperature—20 °C; pH—3 to 4; Time: 40 min;	Mazumdar et al. (2009), Naddeo et al. (2009)
12	Cetirizine dihydrochloride	Ultrasound-assisted enzyme catalyzed	Power—100 W; Frequency—25 kHz Time—7 h; Temperature—50 °C	Tran et al. (2015), Maji et al. (2017)
13	Fluoxetine	Sono-biological	Power—60 W; Frequency—60 kHz; Temperature—20 °C; microorganisms used in biological processes	Karine de Sousa et al. (2018)
14	Cefixime	Photocatalysis	Temperature—25 °C pH—3 to 11, time 15–90 min [H$_2$O$_2$]—0.05 to 0.85 mL/L	Hasani et al. (2020)
15	Rifampicin	Ultrasonic horn	Temperature—60 min, pH—natural, time—60 min	Afroozān Bāzghale and Mohammad-Khāh (2020)
16	Bisphenol S		Frequency—620 kHz; power—80 W pH—4; T—23 ± 1 °C; 0.1 mM of persulfate; reaction Time—60 min	Nejumal et al. (2023)

(continued)

4.3 Elimination of Hazardous Substances

Table 4.1 (continued)

Sr. No.	Pollutant	Sonochemical method	Conditions for degradation	References
17	Sulfamethoxazole	Ultrasonic horn	Temperature—25 °C pH—3 to 11, time—120 min	Moradi et al. (2020)
18	Tylosin	Ultrasonic bath	Temperature—10 to 40 °C; pH—3 to 11; Time—10 to 120 min	Yousef Tizhoosh et al. (2020)

4.4 The Use of Less Hazardous Chemicals and Environmentally Friendly Solvents

In any chemical synthesis, solvent-free reactions have a more significant edge than those involving solvent due to their green approach and easier workup. Therefore, a range of solvent-less organic processes can be carried out by avoiding conventional volatile organic solvents and using techniques such as microwave, ultrasound, UV–visible, infrared radiation, grinding, and milling. By reducing waste product output, reaction time, and energy, these greener approaches make solvent-free reactions more cost-effective and environmentally beneficial. Numerous physiologically active compounds, including coumarin, flunixin, benzimidazole moiety, benzodiazepines, barbituric acid analogs, and chalcones, are produced using solvent-less chemical synthesis (Shaoo and Banik 2024).

The use of a catalyst (nano-based) is one of the principles used in the green chemistry approach. Using solid supports and suitable solvents, catalysts can be doped to stabilize nanostructured materials, particularly metal colloids (less than 10 nm). The US is superior in creating metal oxides and other nanomaterials by facilitating the production of nano-sized catalysts and amorphous particles. A commonly reported example of this would be the production of noble metal particles and bimetallic nanoparticles in aqueous solutions using sonochemistry. An approach for preparing palladium NPs has been reported in the late 90s. Then, it reduces an aqueous solution of tetrachloropalladate with an organic addition to produce nanoparticles with interstitial carbon (PdCx). This latter acts as a catalyst for Pd(II) reduction. The reaction proceeds via the Scheme 4.1 (Cintas and Luche 1999).

In another study conducted by Massimiliano Lupacchini et al. advanced composites were synthesized using a sonochemical approach, which used ultrasonic energy to improve the stability and characteristics of the material. Ultrasonic equipment was used at a high frequency of 475 kHz and an intensity of 50 W/cm^2. Silver nitrate and anilinium nitrate were sonicated in an aqueous solution at room temperature to produce a silver

$$H_2O \longrightarrow H^\bullet + HO^\bullet$$
$$R\text{-}H + H^\bullet/HO^\bullet \longrightarrow R^\bullet + H_2O/H_2$$
$$R\text{-}H \longrightarrow \text{Pyrolysis radicals}$$
$$Pd(II) + \text{Reducing radicals} \longrightarrow Pd(0)$$
$$mPd(0) \longrightarrow Pd_m \text{ (active)}$$
$$Pd_m \text{ (active)} + RH \longrightarrow Pd_m\text{-RH (adsorption)}$$
$$Pd_m\text{-RH} \longrightarrow (PdC)_m \longrightarrow (PdC)_n \text{ aggregation}$$

Reaction Scheme 4.1 Mechanism of synthesis of NPs using sonochemistry nanowire/PANI (emeraldine salt) composite. Isopropyl alcohol was used for dual benefits: (i) for scavenging generated OH radicals and (ii) for enhancing the characteristics and dispersion stability of the silver wire. The scavenging rate of aniline was reduced, and hence, the process was more regulated (Lupacchini et al. 2017).

Ionic liquids (ILs) are becoming increasingly popular as substitute solvents in sonochemistry because of their unique qualities, including high density, heat capacity, viscosity, and vapor pressure. These properties help improve reaction efficiency in viscous media by increasing acoustic energy absorption inside collapsing cavities. Furthermore, Mass transfer and mixing of reactants increases due to US. ILs are convenient as a reaction medium or as reactants as they have simplified workup processes and easy recycling. For use in rechargeable batteries, magnetic devices, catalysis, and other fields, nickel sulfide in the micron form is essential. A novel synthetic method was described for the sonochemical synthesis of NiS submicron particles in 1-butyl-3-methylimidazolium tetrafluoroborate [BMIM][BF4]. A Ti horn with a diameter of 0.635 cm was used to expose the reaction mixture to 20 kHz and 100 W of radiation, which was generated by incorporating the precursor into a mixture of solvents (ethanol with IL in an 8:2 ratio) (Lupacchini et al. 2017).

Several studies suggest nanoparticle-based antibiotics may improve their effectiveness against resistant and nonresistant bacteria. An experiment was conducted to study penicillin's effect in its nanoparticulate form. Powdered penicillin was suspended in deionized water to produce the NPs particles, which were then sonicated for 10 min at 20 kHz to boost the antibiotic's effectiveness. The sample's minimum inhibitory concentration (MIC) was determined by testing it against *Staphylococcus aureus*. The MIC for the sonicated sample increased by two times compared to its bulk form. The required quantity of paracetamol was 0.2 mg/mL compared, whereas 33 mg/mL was needed in the coarse form. This technique might offer an innovative approach to extending the useful life of antibiotics that are no longer effective in their conventional formulation. Smaller treatment doses are possible due to the increase in efficiency, which would lessen both the antibiotics' adverse effects and environmental waste (Ellstrom and Török 2018).

Jin et al. established "chemical aerosol flow synthesis" to prepare quantum dots. This approach replaced organic solutions of high densities (like octadecane) with nanoparticle precursors for aqueous-based systems. The organic incorporated solutions were diluted with a lower viscous solvent to reduce their time for ultrasonic nebulization. A concentrated precursor solution remains in the high-density solvent after the vaporized droplets of the solution pass through a heating zone and initially lose the low boiling point solvent. Higher-density organic liquids undergo chemical reactions to produce crystal-like nanoparticles, which are then cooled in evaporators with cool solvents. In particular, fluorescent Cd-based quantum dots, including mixed chalcogenides and CdS, CdSe, and CdTe, have been successfully prepared using this technique. By adjusting the furnace temperature, particle surface area can be easily adjusted along with the photoluminescent emission of these dots. With this method, one droplet can produce hundreds of NPs, and high-density liquids can stop the NPs from clumping together. A highly affordable, gram-scale synthesis was also achievable, which had been a challenge with traditional quantum dot syntheses. Even at a small-scale setup in a laboratory, the production rate was approximately 100 mg/h (Bang and Suslick 2010).

4.5 Environmental Remediation

As we have seen, pollutants emerging from industries (chemical, textile, pharma), household waste, and hospital waste cause significant harm to the environment. Their treatment through conventional methods is not sufficient for their complete degradation. Hence, the sonochemical approach is crucial. Several published articles discuss the involvement of sonochemistry in the degradation of substances that cause harm to the environment. The research focuses on the modifications in sonochemical reactors, their time, power duration, byproducts, and various other factors for efficient degradation (Rosales Pérez and Esquivel Escalante 2024).

4.5.1 Sonochemistry in Wastewater Treatment

Wastewater treatment is one of the main challenges facing the entire world today. Many countries adopt to recycle the WW emerging from chemical industries after preliminary treatment. However, doing so is not always safe and convenient, as it causes numerous adverse effects. This water can damage aquatic life and can also prove to be fatal to humans. A vast amount of research is currently being done on WWT. Some of them are presented below (Mahamuni and Adewuyi 2010).

Linear alkylbenzene sulphonate (LAS) is a surfactant commonly discharged from the textile industry effluent. Dehgani et al. have reported the usage of a sonochemical reactor for the decomposition of LAS. They used a methylene blue active substance as a reference to understand the effect of LAS degradation. The frequency required is more than 10 kHz at a high power of 400 W with a neutral pH at normal temperature. Time and power are the crucial factors for LAS degradation (Dehghani et al. 2010).

As discussed earlier, the pharmaceutical industry also contributes significantly to wastewater. The major problem is caused by anti-infective agents, as these agents, if consumed unnecessarily, increase the chances of resistance, thereby increasing health problems in urban areas. Sonolysis is one of the emerging processes for degrading antibiotics from wastewater. Low antibiotic concentrations that are typically in nanomolar quantities can be degraded with less US frequency (< 250 kHz) and slightly moderate temperature ranges. The sonochemical reactors are designed accordingly for efficient

4.5 Environmental Remediation

degradation. This method is suitable for its efficiency, less time, and basic operation mechanism. However, its scale-up to large concentrations of antibiotics is still a challenge (Pirsaheb et al. 2023).

Heavy metals that majorly arise from the steel, battery, and electrical industries are some of the most fatal inorganic pollutants in wastewater. Surprisingly, their nanoparticles are effective candidates for the treatment of wastewater. Yadav et al. have reported one such method of synthesizing iron oxide nanoparticles (IONP) for WWT. The NPs are characterized by various analytical techniques and are used to absorb lead and cadmium metals from the fly ash solutions. Fly ash is emitted from cement manufacturing units, biofertilizers, and other construction sites. Nearly 0.6 mg of IONPs in a decaliter of water removed Cd and Pb from the fly ash solutions with more than 95% efficiency. Further, the IONPs are recyclable and can be recovered by applying the magnetic field, making them one of the greenest and most budget-friendly approaches for heavy metal removal (Yadav et al. 2020).

Similar to the previous case study, sonochemistry combined with NP systems work wonders in WWT. As discussed in the earlier chapters, sonochemical processes generate several radicals and oxidants; iron NPs are used to absorb these oxidants. Iron, in its ferric form, forms links with these radicals and accelerates their reactions by oxidizing them. Furthermore, they adsorb the heavy metals and other inorganic materials present in WW, serving a dual purpose. Further, a novel catalyst in which copper combined with Fe NPs is prepared for the degradation of Rhodamine. Similarly, silver and FE NPs are reported for degradation of methylene blue. Both catalysts are prepared using hydrothermal reactors in the presence of DMF as a solvent. The catalysts are characterized by various analytical methods for their size, shape, and other structural properties. The degradation rate of the pollutants these NPs mentioned above is roughly 95% within 1–1.5 h, making them highly preferred (Bao et al. 2023).

As the wastewater emerging from industries concerns the environment, wastewater from households and complexes is equally tedious. Sewage treatment is one of the significant challenges municipal corporations face in cities. Many corporations conduct primary and secondary treatment of sludge; however, due to the hard treatment of sludge, the growth of plantations in nearby areas is affected. Hence, other robust alternatives are necessary. US degradation of sludge proceeds in three primary mechanisms: (i) free radical oxidation, (ii) shearing effect, and (iii) degradation in the presence of heat. The first mechanisms generate hydroxy and hydrogen radicals, 2nd and 3rd break down the sludge. Further, the efficiency increases due to hydrogen peroxide. All of these processes proceed in the power range of 200 to 1000 kHz (Djellabi et al. 2024).

4.5.2 Sonochemistry in Sludge Treatment

Sludge dewatering is another essential process, like sludge treatment, and the US has also found its application in this domain. It was observed that the US has the potential to split the sludge flocks and liberate the water molecules associated with them. This releases several inorganic ions and carbon trioxides. Due to this, the pH and heat generated by the system increase. The optimal conditions for achieving this are a power of 9.8 W and the time required is half a second. A higher amount of energy or ultrasonic frequency may cause undesirable outcomes; hence, properly setting all the parameters becomes crucial (Qi et al. 2024).

As evident from the previous case studies discussed, sludge treatment is of utmost importance as sludge and activated sludge increase the yearly waste and pose a risk to the biotic life. In the last decade, the US and other treatment processes have emerged as a robust alternative to sludge treatment. A recent case study describes that standalone microwave irradiation surges the carbon-based content in the soluble phase, leading to better decomposition. Further, if the medium is made alkaline with metal hydroxides, the decomposition rate increases by almost 60%, along with increased biogas yield. However, further optimization, using microwaves to remove more organic content, and pilot-scale studies are essential for sludge treatment (Jákói et al. 2021). Further, the use of the US in soil treatment has been discussed in Sect. 4.8.

4.6 Pollution Prevention

As evident from the previously discussed points, sonochemistry is one method that contributes to the environment. Recently, researchers have shown interest in research in which semiconductor materials, including metal oxides, graphenes, graphite, etc., have been used as photocatalysts for degrading organic pollutants. The US, along with photocatalysts, also called sonophotolytic processes, is also of recent interest as researchers believe combining these processes would enhance the degradation of pollutants (Theerthagiri et al. 2021). The increase in the quantity of carbon dioxide has increased in recent years in the atmosphere and has been a cause of the melting of glaciers at the poles and the depletion of the ozone layer, thus making it the central region contributing to global warming. US enhances photocatalysis through advanced synthesis and reaction techniques. In sonochemical synthesis, ultrasound creates extreme conditions that help form high-quality photocatalysts with improved nanostructures, such as well-dispersed nanoparticles and defect-rich materials. The US also boosts photocatalytic processes by improving the movement of reactants, creating reactive radicals, and keeping catalyst surfaces active. Additionally, piezoelectric materials generate electrical charges under ultrasound, improving light absorption and charge separation, which enhances photocatalytic efficiency. These methods are applied in areas like pollutant removal and hydrogen production,

4.6 Pollution Prevention

offering improved performance, lower energy requirements, and better catalyst reusability. Optimizing factors like ultrasound intensity and catalyst dosage further optimizes these processes (Meroni and Bianchi 2022). NPs and piezocatalysis have been reported for their reduced CO_2 efficiency. $BaTiO_2$ nanofibers at a concentration of 12,525 µmol/g/h under the sonochemical conditions caused a reduction in CO_2 reduction rate with a selectivity rate of above 90% (Kerboua 2024).

Trapping CO_2 from the atmosphere and then converting it into hydrocarbons using sonochemical approaches has been an area of interest for most of the researchers working in this domain. The Sabatier reaction is the most important application of the conversion of CO_2 into green sources. In this reaction, hydrogen plays important role for conversion of CO_2 to methane. Hydrogen acts as a donor for reduction of CO_2 to CO in the reaction, along with scavenging the generated hydroxy radicals for conversion to water. Simple hydrocarbons such as methane, ethane, and alkylene are formed by this reaction. If argon is added to an equal amount of hydrogen, it catalyzes the reaction rapidly. Further addition of sodium chloride enhances the process (Abdin and Khalilpour 2019; Dehane et al. 2022) (Reaction Scheme 4.2).

Reducing nitrogen to ammonia with the use of the US has also emerged these days. A recent study synthesized nitrogen-doped MoS_2 ($N-MoS_2$) photocatalyst (NMPC) by a one-step protocol forming nanosheets. Platinum NPs were added to NMPC via the photo-ultrasonic reduction method. This combination was found to boost nitrogen fixation, and the researchers concluded that the efficiency of this system was increased due to nitrogen doping. Nitrogen fixation was seen both under UV and visible light with a rate of 133.8 µmol/g(cat)·h. This rate is significantly higher when compared with either sonocatalysis or photocatalysis (Maimaitizi et al. 2020).

The successful synthesis of nitrogen-doped graphene/porous $g-C_3N_4$ composites involved the vibration of porous $g-C_3N_4$, triazine, CaC_2 and ethanol with the aid of ultrasound. Various methods, such as spectroscopy, X-ray diffraction, and electron microscopy, were employed to examine the materials' composition, structure, and characteristics. The N-doped graphyne in the composites is firmly bonded to $g-C_3N_4$ flakes by a flexible, folded lamellar structure. These composites significantly improved photocatalytic performance, achieving up to 1.68–1.44 times better degradation of pollutants like methylene blue, rhodamine B, tetracycline and levofloxacin compared to pure $g-C_3N_4$. They also showed a remarkable increase in ammonia production, reaching 1.71 mmol g^{-1} h^{-1}, nearly six times higher than $g-C_3N_4$ alone. Various electrical and optical analyses confirmed that enhanced photocatalytic activity was attributed to better light absorption and improved charge separation. This study demonstrates that N-doped graphyne is an effective modifier for $g-C_3N_4$ and may help improve other photocatalysts (Zhao et al. 2024).

$$CO_2 + 4H_2 \longrightarrow CH_4 + 2H_2O$$

Reaction Scheme 4.2 Sabatier reaction

4.7 Purification of Water

Water is utmost necessary for all biological lives. It is a well-known fact that although 75% of the Earth contains water and 98% is present in seas, only 2% of water remains fresh water for humans. Of this 2%, only 0.036% can be used as potable drinking water, while the remaining is found in glaciers and underground water sources. Due to the increase in water pollution, the threat of polluting these essential water sources looms large. Hence, there is a potential need for adequate water purification to meet the needs of humans (Srivastava and Yadav 2023).

US is widely used for WWT, as described earlier. However, it can work wonders with the purification of water. One such approach is membrane filtration. It is widely reported that the US enhances membrane filtration by improving water movement through both ceramic and organic membranes. This process accumulates particles on the membrane surface, but the increased mass transfer rate primarily causes it. Traditionally, membrane separation involved the formation of a fouling layer on the membrane surface, which reduced process efficiency. Combined with the US-assisted forward flushing of the membrane, the membrane was restored, and its life was extended. Nevertheless, this technology was used in the initial 20s. Advanced processes and reactors are available and constantly needed for scale-up (Fetyan and Salem Attia 2020).

In a recent study, functionalized polyelectrolytes were synthesized on a single-wall carbon nanotube (SWCNT) using sonochemically catalyzed atom transfer radical polymerization (SONO-ATRP). The study was conducted in water for the formation of nanoresins, and it was found that minimal catalysts were used to avoid any initiator or reducing agent in the process. The formed nanoresins performed exceptionally well in water purification due to their high absorbance capacity during normal temperature and pressure conditions. SONO-ATRP of vinyl benzyl trimethyl ammonium chloride (vbTMAC) demonstrated good monomer conversion at a frequency of 20 kHz for 240 min at RT. The nanoresins also showed more incredible Thomas Model breakthrough curves when compared to the surrogate analyte. Another advantage of these materials is that they are easily recyclable (Sahu et al. 2020).

In brief, it can be concluded that most researchers have opted for different reactors for water purification combined with the US (Fig. 4.1) and have also tried to scale up. However, its application in wastewater treatment and further treatment for its purification is a challenge. Further, microflow sonoreactors are a good alternative for their low-energy consumption. However, they have the disadvantage that if scaled up, it may cause clogging due to solid particulate matter. Further, the researchers also address the use of continuous flow sonochemistry to address the issue. In the long term, the economics of this process reduce the cost significantly due to lower energy consumption. However, the literature is mainly available theoretically, and efficient modeling is required to carry it up to higher scales (Kerboua and Hamdaoui 2021).

4.8 Decontamination of Soil

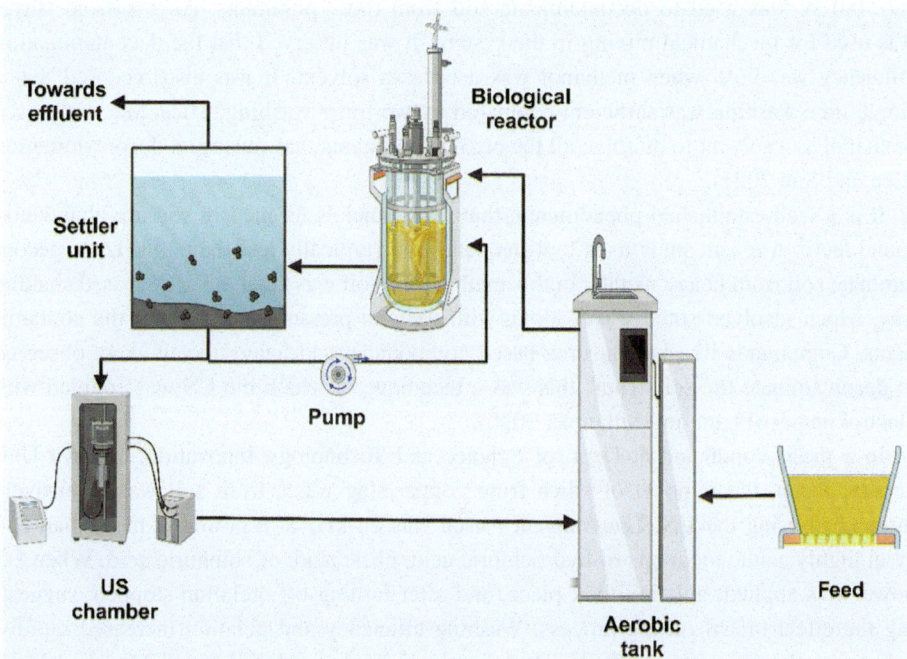

Fig. 4.1 US combined with wastewater treatment plant (created using Biorender.com)

4.8 Decontamination of Soil

Soil is essential for humans as it is directly related to agricultural activities, mining, plantations, and building infrastructure. Due to leaching from underground water sources, landfilling, and other and other associated activities, soil pollution is increasing rapidly. When contaminated with pollutants, these pollutants can enter human bodies directly through vegetation. Hence, the removal of contaminants from the soil is of utmost necessity (Menon 2023).

In a study on a laboratory scale, a soil sample (sludge) was taken, which had a density of 1.3 g/mL. The sample was dried, passed through a 1.18 mm pore-size sieve, and then exposed to a US bath with seven transducers ($42 \times 30 \times 10$ cm^3). A frequency of 28 kHz was given to the system with a wide range of intensities extending from 350 to 1200 W. The primary aim of the experiment was to wash out the soil contaminants in acidic and basic media using a US application. After the treatment, the mixture was filtered and dried at 35 °C. The dried extract was pressed into pellets and analyzed for soil content by an XRF analyzer. The US was determined to be effective for soil purification (Tri 2023).

Most organic pollutants belong to a class of non-polar organic pollutants commonly found in soil. In a study conducted, a double bath US system at 28 kHz US transducer and 480 W was used to decontaminate soil from these pollutants. An overhead stirrer was used for mechanical mixing in the system. It was observed that the decontamination efficiency was 90% when methanol was used as a solvent. It was also recorded that a single-time washing was sufficient compared to two times washing. Thus, this study used methanol as a solvent to dissolve all the organic pollutants that cause soil decontamination (Lee and Son 2021).

It is a well-established phenomenon that heavy metals leached in soil are abundantly found there. A recent study used low-frequency US (typically less than 20 kHz) to decontaminate soil from heavy metals. In this method, the soil was taken on a US-based shaking tray, which involved sonic waves along with vacuum pressure to leach out the contaminants. Compounds like kaolin, urea-based compounds, and heavy metals were observed to decontaminate the soil. Thus, this was a technique in which the US was coupled with electrokinetics (Pham and Sillanpää 2020).

In a thesis conducted at Dept. of Science and Technology Innovation, Nagoya University, Japan, the removal of silica from copper slag waste from soil was extensively researched using the US. The frequency used was 26 kHz at a power of more than 550 W at highly acidic mediums of hydrochloric acid, nitric acid, or sulphuric acid. When US power was applied, gelation took place, and after turning off, gelation stopped, suggesting the effect of US on the process. Washing efficiency and gelation increased rapidly, increasing the power to 1200 W. The particle size was analyzed by SEM, which gave satisfactory results (Tri 2023).

Uranium abundantly exists as a soil pollutant and adversely affects soil purity. In a study, a 2^3 factorial design was used to derive experiments, which included mechanical washing combined with US to wash off uranium. The varying factors were the agitation speed, power, frequency, temperature, time, etc. It was observed that the US enhanced the washing process by 1/4th time compared to regular mechanical washing, reducing the time required from 2 h to just 30 min. Genetic algorithms were used to optimize the process, and it was found that the optimized conditions had a temperature of 50 °C at a slightly alkaline pH and a frequency of 24 kHz. The removal efficiency was above 80% for soil with high organic matter and clay content and more than 65% for soil deprived of clay and organic matter (Radu et al. 2020).

4.9 Conclusion

Environmental pollution has been an increasing cause of concern for biotic and abiotic life for ages. Human lifestyle changes, rapidly growing industries, and the migration of people toward metro cities are the critical factors driving this. It is the need of the hour to take good remedies against this growing concern. Using ultrasound is reported to degrade pollutants found in the waterbodies, soil and atmosphere. Further, ultrasound generates environmentally safe products, thus proving to be a green technology. This chapter highlights the necessary case studies that have been proven to be effective against pollutants found in the environment.

References

Abdin Z, Khalilpour KR (2019) Single and polystorage technologies for renewable-based hybrid energy systems. In: Polygeneration with polystorage for chemical and energy hubs. Elsevier, pp 77–131. https://doi.org/10.1016/B978-0-12-813306-4.00004-5

Adewuyi YG (2001) Sonochemistry: environmental science and engineering applications. Ind Eng Chem Res 40(22):4681–4715. https://doi.org/10.1021/ie010096l

Adewu-Yi YG (2005) Sonochemistry in environmental remediation. 2. Heterogeneous sonophotocatalytic oxidation processes for the treatment of pollutants in water. Environ Sci Technol 39(22):8557–8570. https://doi.org/10.1021/es0509127

Afroozān Bāzghale Ā, Mohammad-Khāh A (2020) Improvement of ultrasound-assisted removal of rifampin in the presence of N: ZnO/GO nanocomposite as sonocatalyst. ChemistrySelect 5(15):4413–4421. https://doi.org/10.1002/slct.202000068

Akanyange SN et al (2021) Does microplastic really represent a threat? A review of the atmospheric contamination sources and potential impacts. Sci Tot Environ 777:146020. https://doi.org/10.1016/j.scitotenv.2021.146020

Akiya N, Savage PE (2002) Roles of water for chemical reactions in high-temperature water. Chem Rev 102(8):2725–2750. https://doi.org/10.1021/cr000668w

Al-Rubaiey NA (2024) Trends in sonochemical treatment of oily wastewater. Petroleum Chemistry. https://doi.org/10.1134/S096554412407003X

Andreozzi R, Raffaele M, Nicklas P (2003) Pharmaceuticals in STP effluents and their solar photodegradation in aquatic environment. Chemosphere 50(10):1319–1330. https://doi.org/10.1016/S0045-6535(02)00769-5

Bang JH, Suslick KS (2010) Applications of ultrasound to the synthesis of nanostructured materials. Adv Mater 22(10):1039–1059. https://doi.org/10.1002/adma.200904093

Bao J et al (2023) Sonoactivated nanomaterials: a potent armament for wastewater treatment. Ultrason Sonochem 99:106569. https://doi.org/10.1016/j.ultsonch.2023.106569

Checa-Fernández A et al (2022) Enhanced remediation of a real HCH-polluted soil by the synergetic alkaline and ultrasonic activation of persulfate. Chem Eng J 440:135901. https://doi.org/10.1016/j.cej.2022.135901

Cintas P, Luche J-L (1999) Green chemistry. Green Chem 1(3):115–125. https://doi.org/10.1039/a900593e

Colarusso P, Serpone N (1996) Environmental detoxification. Interpretation J Bible Theol 22(1):61–89

Crini G, Lichtfouse E (2019) Advantages and disadvantages of techniques used for wastewater treatment. Environ Chem Lett 17(1):145–155. https://doi.org/10.1007/s10311-018-0785-9

Davies JF et al (2014) Temperature dependence of the vapor pressure and evaporation coefficient of supercooled water. J Geophys Res: Atmospheres 119(18):10931–10940. https://doi.org/10.1002/2014JD022093

Dehane A et al (2022) Sonochemical and sono-assisted reduction of carbon dioxide: a critical review. Chem Eng Process—Process Intensification 179:109075. https://doi.org/10.1016/j.cep.2022.109075

Dehghani MH, Najafpoor AA, Azam K (2010) Using sonochemical reactor for degradation of LAS from effluent of wastewater treatment plant. Desalination 250(1):82–86. https://doi.org/10.1016/j.desal.2009.05.011

Djellabi R et al (2024) Ultrasonic disintegration of municipal sludge: fundamental mechanisms, process intensification and industrial sono-reactors. ChemPlusChem. https://doi.org/10.1002/cplu.202400016

Ellstrom CJ, Török B (2018) Application of sonochemical activation in green synthesis. In: Green chemistry. Elsevier, pp 673–693. https://doi.org/10.1016/B978-0-12-809270-5.00024-8

Fedorov K et al (2024) High-performance activation of ozone by sonocavitation for BTEX degradation in water. J Environ Manage 363:121343. https://doi.org/10.1016/j.jenvman.2024.121343

Fetyan NAH, Salem Attia TM (2020) Water purification using ultrasound waves: application and challenges. Arab J Basic Appl Sci 27(1):194–207. https://doi.org/10.1080/25765299.2020.1762294

Gupta P et al (2021) Sonochemical degradation of polycyclic aromatic hydrocarbons: a review. Environ Chem Lett 19(3):2663–2687. https://doi.org/10.1007/s10311-020-01157-9

Halling-Sørensen B et al (1998) Occurrence, fate and effects of pharmaceutical substances in the environment—a review. Chemosphere 36(2):357–393. https://doi.org/10.1016/S0045-6535(97)00354-8

Hasani K et al (2020) The efficacy of sono-electro-Fenton process for removal of Cefixime antibiotic from aqueous solutions by response surface methodology (RSM) and evaluation of toxicity of effluent by microorganisms. Arab J Chem 13(7):6122–6139. https://doi.org/10.1016/j.arabjc.2020.05.012

Hoang AQ et al (2023) Polycyclic aromatic hydrocarbons in air and dust samples from Vietnamese end-of-life vehicle processing workshops: contamination status, sources, and exposure risks. Bull Environ Contam Toxicol 110(6):110. https://doi.org/10.1007/s00128-023-03757-x

Hu D, Liu S, Zhang G (2025) Sonochemical treatment for removal of aqueous organic pollutants: principles, overview and prospects. Sep Purif Technol 353:128264. https://doi.org/10.1016/j.seppur.2024.128264

Ince NH et al (2001) Ultrasound as a catalyzer of aqueous reaction systems: the state of the art and environmental applications. Appl Catal B 29(3):167–176. https://doi.org/10.1016/S0926-3373(00)00224-1

Jákói Z et al (2021) Microwave and ultrasound based methods in sludge treatment: a review. Appl Sci 11(15):7067. https://doi.org/10.3390/app11157067

Karine de Sousa A et al (2018) New roles of fluoxetine in pharmacology: antibacterial effect and modulation of antibiotic activity'. Microb Pathog 123:368–371. https://doi.org/10.1016/j.micpath.2018.07.040

Kerboua K (2024) Chapter 14 The sonochemical reduction of carbon dioxide. In: Sonochemical water and wastewater decontamination. De Gruyter, pp 339–354. https://doi.org/10.1515/9783111137940-014

Kerboua K, Hamdaoui O (2021) Sonochemistry for water remediation: toward an up-scaled continuous technology. In: Applied water science. Wiley, pp 437–467. https://doi.org/10.1002/9781119725282.ch13

Khan MT et al (2021) Hospital wastewater as a source of environmental contamination: an overview of management practices, environmental risks, and treatment processes. J Water Process Eng 41:101990. https://doi.org/10.1016/j.jwpe.2021.101990

Kumar P et al (2021) Biological contaminants in the indoor air environment and their impacts on human health. Air Qual Atmos Health 14(11):1723–1736. https://doi.org/10.1007/s11869-021-00978-z

Kummerer K (2003) Significance of antibiotics in the environment. J Antimicrob Chemother 52(1):5–7. https://doi.org/10.1093/jac/dkg293

Lee D, Son Y (2021) Ultrasound-assisted soil washing processes using organic solvents for the remediation of PCBs-contaminated soils. Ultrason Sonochem 80:105825. https://doi.org/10.1016/j.ultsonch.2021.105825

Lupacchini M et al (2017) Sonochemistry in non-conventional, green solvents or solvent-free reactions. Tetrahedron 73(6):609–653. https://doi.org/10.1016/j.tet.2016.12.014

Mahamuni NN, Adewuyi YG (2010) Advanced oxidation processes (AOPs) involving ultrasound for waste water treatment: a review with emphasis on cost estimation. Ultrason Sonochem 17(6):990–1003. https://doi.org/10.1016/j.ultsonch.2009.09.005

Maimaitizi H et al (2020) Facile photo-ultrasonic assisted synthesis of flower-like Pt/N-MoS$_2$ microsphere as an efficient sonophotocatalyst for nitrogen fixation. Ultrason Sonochem 63:104956. https://doi.org/10.1016/j.ultsonch.2019.104956

Maji HS, Maji S, Bhattacharya M (2017) An exploratory study on the antimicrobial activity of cetirizine dihydrochloride. Indian J Pharm Sci 79(5). https://doi.org/10.4172/pharmaceutical-sciences.1000288

Mason TJ (1999) Sonochemistry: current uses and future prospects in the chemical and processing industries. Philos Trans R Soc London. Ser A: Math Phys Eng Sci 357(1751):355–369. https://doi.org/10.1098/rsta.1999.0331

Mason TJ (2007) Sonochemistry and the environment—providing a "green" link between chemistry, physics and engineering. Ultrason Sonochem 14(4):476–483. https://doi.org/10.1016/j.ultsonch.2006.10.008

Mazumdar K et al (2009) The anti-inflammatory non-antibiotic helper compound diclofenac: an antibacterial drug target. Eur J Clin Microbiol Infect Dis 28(8):881–891. https://doi.org/10.1007/s10096-009-0739-z

Menon R (2023) Soil degradation in India spells doom for millions. Available at: https://india.mongabay.com/2023/10/soil-degradation-in-india-spells-doom-for-millions/

Meroni D, Bianchi CL (2022) Ultrasound waves at the service of photocatalysis: from sonochemical synthesis to ultrasound-assisted and piezo-enhanced photocatalysis. Curr Opin Green Sustain Chem 36:100639. https://doi.org/10.1016/j.cogsc.2022.100639

Moradi M et al (2020) A review on pollutants removal by Sono-photo-Fenton processes. J Environ Chem Eng 8(5):104330. https://doi.org/10.1016/j.jece.2020.104330

Naddeo V et al (2009) Degradation of diclofenac during sonolysis, ozonation and their simultaneous application. Ultrason Sonochem 16(6):790–794. https://doi.org/10.1016/j.ultsonch.2009.03.003

Naddeo V et al (2012) Degradation of antibiotics in wastewater during sonolysis, ozonation, and their simultaneous application: operating conditions effects and processes evaluation. Int J Photoenergy 2012:1–7. https://doi.org/10.1155/2012/624270

Nejumal KK et al (2023) Degradation studies of bisphenol S by ultrasound activated persulfate in aqueous medium. Ultrason Sonochem 101:106700. https://doi.org/10.1016/j.ultsonch.2023.106700

Pham TD, Sillanpää M (2020) Ultrasonic and electrokinetic remediation of low permeability soil contaminated with persistent organic pollutants. In: Advanced water treatment. Elsevier, pp 227–310. https://doi.org/10.1016/B978-0-12-819227-6.00004-8

Pirsaheb M, Moradi N, Hossini H (2023) Sonochemical processes for antibiotics removal from water and wastewater: a systematic review. Chem Eng Res des 189:401–439. https://doi.org/10.1016/j.cherd.2022.11.019

Qi Y et al (2024) Optimizing sludge dewatering efficiency with ultrasonic treatment: insights into parameters effects, and microstructural changes. Ultrasonics Sonochem 102:106736. https://doi.org/10.1016/j.ultsonch.2023.106736

Radu DA et al (2020) Optimization of uranium soil decontamination in alkaline washing using mechanical stirring and ultrasound field. Environ Sci Pollut Res 27(6):5941–5950. https://doi.org/10.1007/s11356-019-07063-0

Rawat S, Singh J, Koduru JR (2021) Effect of ultrasonic waves on degradation of phenol and para-nitrophenol by iron nanoparticles synthesized from Jatropha leaf extract. Environ Technol Innov 24:101857. https://doi.org/10.1016/j.eti.2021.101857

Razavi T et al (2020) Evaluation of the photosonolysis process efficacy for the removal of anionic surfactant linear alkyl benzene sulfonate from aqueous solutions. J Ecol Eng 21(6):1–7. https://doi.org/10.12911/22998993/122192

Rosales Pérez A, Esquivel Escalante K (2024) The evolution of sonochemistry: from the beginnings to novel applications. ChemPlusChem 89(6). https://doi.org/10.1002/cplu.202300660

Sahu A, Sheikh R, Poler JC (2020) Green sonochemical synthesis, kinetics and functionalization of nanoscale anion exchange resins and their performance as water purification membranes. Ultrason Sonochem 67:105163. https://doi.org/10.1016/j.ultsonch.2020.105163

Salgot M, Folch M (2018) Wastewater treatment and water reuse. Curr Opin Environ Sci Health 2:64–74. https://doi.org/10.1016/j.coesh.2018.03.005

Sathishkumar P, Viswanathan R (2016) Review on the recent improvements in sonochemical and combined sonochemical oxidation processes—a powerful tool for destruction of environmental contaminants. Renew Sustain Energy Rev 55:426–454. https://doi.org/10.1016/j.rser.2015.10.139

Savun-Hekimoğlu B (2020) A review on sonochemistry and its environmental applications. Acoustics 2(4):766–775. https://doi.org/10.3390/acoustics2040042

Segura PA et al (2009) Review of the occurrence of anti-infectives in contaminated wastewaters and natural and drinking waters. Environ Health Perspect 117(5):675–684. https://doi.org/10.1289/ehp.11776

Shaoo BM, Banik BK (2024) Solvent-less reactions: green and sustainable approaches in medicinal chemistry. In: Green approaches in medicinal chemistry for sustainable drug design. Elsevier, pp 387–408. https://doi.org/10.1016/B978-0-443-16164-3.00017-0

Sharma G et al (2022) Analysis of industrial pollution in India. Int J Health Sci:2763–2771. https://doi.org/10.53730/ijhs.v6nS5.9242

Srivastava P, Yadav U (2023) Mode of the mechanism of biogenic nanomaterials involved in the adsorption of pollutants. In: Nanobiotechnology for bioremediation. Elsevier, pp 273–296. https://doi.org/10.1016/B978-0-323-91767-4.00009-4

Thanekar P et al (2021) Degradation of benzene present in wastewater using hydrodynamic cavitation in combination with air. Ultrason Sonochem 70:105296. https://doi.org/10.1016/j.ultsonch.2020.105296

Theerthagiri J et al (2021) Application of advanced materials in sonophotocatalytic processes for the remediation of environmental pollutants. J Hazard Mater 412:125245. https://doi.org/10.1016/j.jhazmat.2021.125245

Thomas S et al (2020) Sonochemical degradation of benzenesulfonic acid in aqueous medium. Chemosphere 252:126485. https://doi.org/10.1016/j.chemosphere.2020.126485

Tran N, Drogui P, Brar SK (2015) Sonochemical techniques to degrade pharmaceutical organic pollutants. Environ Chem Lett 13(3):251–268. https://doi.org/10.1007/s10311-015-0512-8

Tri PP (2023) Ultrasonic technology to recover useful components from contaminated soil and industrial waste (September)

Wei Z et al (2021) A review on phytoremediation of contaminants in air, water and soil. J Hazard Mater 403:123658. https://doi.org/10.1016/j.jhazmat.2020.123658

Wood RJ et al (2021) The application of flow to an ultrasonic horn system: phenol degradation and sonoluminescence. Ultrason Sonochem 71:105373. https://doi.org/10.1016/j.ultsonch.2020.105373

Yadav VK et al (2020) Synthesis and characterization of amorphous iron oxide nanoparticles by the sonochemical method and their application for the remediation of heavy metals from wastewater. Nanomaterials 10(8):1551. https://doi.org/10.3390/nano10081551

Yousef Tizhoosh N et al (2020) Ultrasound-engineered synthesis of $WS2@CeO_2$ heterostructure for sonocatalytic degradation of tylosin'. Ultrason Sonochem 67:105114. https://doi.org/10.1016/j.ultsonch.2020.105114

Zhang M et al (2020) Ultrasound-assisted electrochemical treatment for phenolic wastewater'. Ultrason Sonochem 65:105058. https://doi.org/10.1016/j.ultsonch.2020.105058

Zhao J et al (2024) In-situ sonochemical formation of N-graphyne modulated porous $g-C_3N_4$ for boosted photocatalysis degradation of pollutants and nitrogen fixation'. Spectrochim Acta Part A Mol Biomol Spectrosc 320:124629. https://doi.org/10.1016/j.saa.2024.124629

Future Outlook on Sonochemistry

Abstract

Sonochemistry is a new emerging green technology in the current decade. It has tremendous benefits for both humans and the environment. It is quite evident from the previous chapters that it plays a pivotal role in synthesizing nanoparticles in healthcare and environmental remediation and is useful in almost every industry. The number of researchers in this area is increasing day by day. As the book comes to an end, this chapter highlights the scope of sonochemistry in the future. It discusses the scope of sonochemistry in the pharmaceutical domain and its combination with microwaves and gives an overview of the industries currently using sonochemistry.

5.1 Introduction

Sonochemistry is required in every domain, such as pharmaceuticals, petrochemicals, mining, healthcare systems, energy, and other engineering fields. The field is growing tremendously daily and is preferred as it complies with green chemistry. Sonochemical equipment requires less energy, produces byproducts that are safe for the environment, and uses a green approach (Qi et al. 2022). The applications of sonochemistry that are discovered now are in the fields of nanocatalyst, piezoelectric photocatalysis, organic synthesis, analytical methods, etc. (Cairós et al. 2020; Meroni and Bianchi 2022) (Fig. 5.1).

Fig. 5.1 Ever-increasing boon of research in sonochemistry (created using Biorender.com)

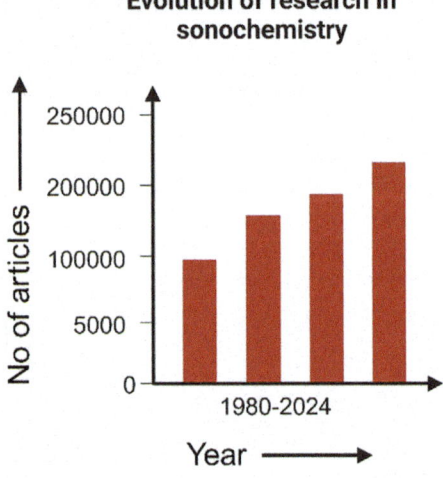

5.2 Sonochemistry and Pharmaceutical Sciences

Sonochemistry is used in pharmaceuticals for the preparation of nanoparticles, their formulation, synthesis of API and their derivatives, and degradation of waste products from pharmaceutical plants.

A sound document example of this is the synthesis of antibiotic tetranosin. The reaction usually takes three complete days using the conventional method. However, sonochemistry drives it in just 10 min. The NaSePH used as a reagent is sonochemically prepared from breaking sodium and PhSeSePh in THf as a solvent. In this way, tetronasin can be synthesized by ultrasound. Like this, sonochemistry can pave other alternatives to synthesize pharmaceutically important agents (Martínez et al. 2021) (Reaction Scheme 5.1).

Sonochemistry is widely used in nanoparticle synthesis, and chitosan NPs are reported to be one of the best materials for controlled drug release. Nanoemulsion of these materials was formulated, and hydrodynamic cavitation was used to enhance the cross-linking of polymers. Different variations, such as the oil used, addition of materials, and viscosity of chitosan, were modified along with other experimental properties. It was observed that good water in oil emulsion was formed using palm oil as an oil phase. The emulsion's particle size was evaluated and found to be appropriate, with a yield of nearly 50%. FTIR analysis shows that the symmetry of palm oil remained unchanged (Zhang et al. 2021).

The development of an electrochemical sensor was achieved using zinc oxide nanoparticles (ZnO NPs) combined with multiwalled carbon nanotubes (MWCNTs) using the US to convert nanocomposites for further detection of epinephrine in formulations as well as human blood samples. The prepared nanocomposites were analyzed using SEM, TEM, FTIR, and other electro-analytical methods. The nanocomposites exhibited significantly

5.2 Sonochemistry and Pharmaceutical Sciences

Reaction Scheme 5.1 Synthesis of tetronasin antibiotic by sonochemistry

enhanced electrocatalytic activity for epinephrine oxidation. The sensor displayed excellent selectivity, reproducibility, and stability, making it suitable for analytical applications. The important kinetic variables, including the number of electrons transferred during the reaction and the heterogeneous rate constant, were evaluated, further confirming the sensor's performance. LOD was found to be 0.016 µM for a range between 0.4 and 2.4 µM, thus providing a technique for detecting epinephrine in formulations and blood samples (Shaikshavali et al. 2020).

In a recent study, a novel approach was performed for synthesizing zero-valent copper (Cu^0) nanoparticles using an ultrasound-assisted method using hibiscus rosa-sinensis extract, which acted as a stabilizing and reducing agent. Applying the sonochemical method to synthesize nanoparticles significantly enhanced the crystalline nature of the nanoparticles and facilitated controlled particle growth. The process leveraged the functional groups in the plant extract, where hydroxyl groups (−OH) acted as reducing agents, converting Cu^{2+} ions into Cu^0. In contrast, carbonyl groups (C=O) from oxidized polyphenols served as capping and stabilizing agents for the nanoparticles. The radicals generated from the sonochemical method facilitated the degradation of pharmaceutical pollutants 5-fluorouracil and lovastatin (> 90%) (Dinesh et al. 2020). In the near future, continuous preparation of these nanoparticles will be possible using ultrasound-assisted approaches. Table 5.1 summarizes the nanoformulation as a pharmaceutical application prepared by sonochemistry.

Table 5.1 List of nanoformulations

S. No.	Name of nanoformulation	Description	References
1	Rapamune® (Sirolimus by Pfizer)	Uses nanocrystals to enhance the solubility and bioavailability of sirolimus	Arti et al. (2022)
2	Diprivan® (Propofol by AstraZeneca)	Nanoemulsion of propofol in a lipid-based system to enhance its solubility and reduce pain at the injection site	Kazi et al. (2023)
3	NutriNano CoQ10® (Coenzyme Q10 by Aquanova)	Micellar curcumin Nanoemulsion increases the bioavailability of CoQ10	Flory et al. (2021)
4	Emend® (Aprepitant by Merck)	An aprepitant nanosuspension, which was found to promote APT's solubility	Liu et al. (2022)

5.3 Sonochemistry and Microwaves

Microwave (MW) falls in the 106 to 109 nm or almost 1 m range in the Infrared spectra. Due to the generated electric field due to microwaves, molecules move back and forth, and this mechanism causes the mixture to heat up. This property is also attributed to their fast penetration into the material, causing changes in their property (Fasina et al. 2003).

MW are typically used for breast cancer detection, and there are two types by which they are detected—tomography and radar-based imaging. The US has also been a method for breast cancer detection. Some researchers are exploring the opportunity to combine these methods for tumor detection. Researchers at Hiroshima University have developed a portable radar-based detector, and the equipment has yielded satisfactory results. The technique is non-invasive, and patient complaints can hence be explored further (AlSawaftah et al. 2022).

Adenomyosis is a widely occurring gynecological disorder in women and is a cause of dysmenorrhea, low iron count, fertility issues, and other reproductive ailments in women. The ultrasound-assisted microwave ablation treatment was given to some patients as a part of clinical trials. A microwave antenna was inserted into adenomyosis under US imaging, and microwave rays have emerged. These rays served as a heat source and caused the lesions' shrinkage by coagulating proteins, thus clearing adenomyosis. The technique is effective as it requires less time, is effective, does not damage other organs, is minimally invasive, and surpasses long-term treatments (Zhang et al. 2022).

Ultrasounds (US) and MW can be used as an alternative for thyroid surgery to remove thyroid nodules. In a clinical study, MW and US were used to remove thyroid nodules, and the success rate was satisfactory. The method is also patient-compliant as it is less invasive and solves cosmetic problems. The only requirement is that the nodule should be benign and requires expert medical professionals. Further research and trials in this

5.4 Industrial Sonochemistry

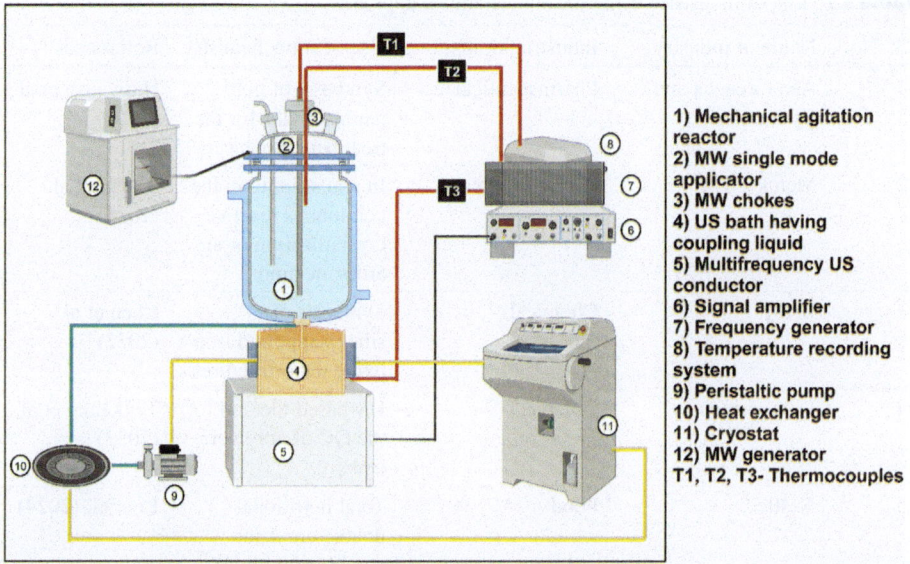

Fig. 5.2 US combined with MW reactor (created using Biorender.com)

domain are necessary for complete validation of the procedure (Gopalakrishnan et al. 2023).

A reactor that combines US and MW (Fig. 5.2) is designed for its intensified effect. The reactor was used to convert *p*-nitrophenol to *p*-nitrocatechol. It was also applicable for the transesterification of Helianthus annuus oil using an acid-based catalyst. The reactor had the advantage of combining the intensities of both US and MW. Further, it could be operated at various ultrasonic frequencies and different power ranges. Further, it could be operated at multimode conditions and can be kept constant as half of the reactor is immersed in a US bath and supplied with a cooling jacket. Figure 5.2 illustrates the use of ultrasounds combined with MW reactors. The authors believe that these types of reactors could be explored for the extraction of thermolabile components in the plants (Gopalakrishnan et al. 2023).

5.4 Industrial Sonochemistry

As it is quite evident from all the information discussed until now that sonochemistry is required in various fields and is environment friendly. Table 5.2 summarizes a list of specific industries that are using sonochemistry. Many more industries may come forward to utilize sonochemical approaches to continuously prepare the various nanoformulations for

Table 5.2 List of industries using the sonochemical approach

S. No.	Name of industry	Industrial domain	Use of sonochemistry	References
1	Astra Zeneca and Cytimmune	Pharmaceutical	Synthesis of gold nanoparticles for the treatment of cancer	Harwansh et al. (2024)
2	Merck	Pharmaceutical	In situ seeding in the continuous reactive crystallization of an aromatic amine	Zhang et al. (2024)
3	BASF	Chemical	Synthesis of silica-supported iron oxide nanostructures	Chen et al. (2022)
4	Dow chemicals	Chemical	US-based method for the QC of polymeric tanks	Tyukanko et al. (2023)
5	Nestle	Food	Total nutritional double emulsion combined with US and HPH	Li et al. (2024)
6	Suez University, project funded by deputyship for research & innovation, ministry of education, Saudi Arabia	Environment safety	Production of vanadium oxide by the US for the removal of organic dye from WW	Ben Aissa et al. (2023)
7	Levi Strauss & Co	Textile	Ultrasound-assisted dyeing of textile	Ben Aissa et al. (2023)
8	Peroxy Chem	Environment	In situ generated hydrogen peroxide for corrosion inhibition, disinfection, and pollutant degradation	Ben Aissa et al. (2023)

pharmaceutical use, treatment wastewater, and combined sonochemistry with microwaves in the upcoming year.

5.5 Conclusion

As this chapter and bookend, the potential uses of sonochemistry in the pharmaceutical field are extensively discussed. The continuous manufacturing of various pharmaceutical APIs and formulations may offer multiple advantages, such as mild conditions, operational simplicity, greenness of the processes, and environmental friendliness. The field finds future directions in the pharmaceutical domains and can be combined with other advanced analytical approaches for more effectiveness. In addition, this branch of science is helpful in almost all domains and needs of the twenty-first century.

References

AlSawaftah N et al (2022) Microwave imaging for early breast cancer detection: current state, challenges, and future directions. J Imaging 8(5):123. https://doi.org/10.3390/jimaging8050123

Arti S et al (2022) Drug nanocrystals as nanocarrier-based drug delivery systems. In: Industrial applications of nanocrystals. Elsevier, pp 179–203. https://doi.org/10.1016/B978-0-12-824024-3.00018-X

Ben Aissa MA et al (2023) Dependency of crystal violet dye removal behaviors onto mesoporous V_2O_5-g-C_3N_4 constructed by simplistic ultrasonic method. Inorganics 11(4):146. https://doi.org/10.3390/inorganics11040146

Cairós C, González-Sálamo J, Hernández-Borges J (2020) The current binomial sonochemistry-analytical chemistry. J Chromatogr A 1614:460511. https://doi.org/10.1016/j.chroma.2019.460511

Chen L et al (2022) Sonochemical synthesis of silica-supported iron oxide nanostructures and their application as catalysts in Fischer-Tropsch synthesis. Micro 2(4):632–648. https://doi.org/10.3390/micro2040042

Dinesh GK, Pramod M, Chakma S (2020) Sonochemical synthesis of amphoteric Cu0-Nanoparticles using Hibiscus rosa-sinensis extract and their applications for degradation of 5-fluorouracil and lovastatin drugs. J Hazard Mater 399:123035. https://doi.org/10.1016/j.jhazmat.2020.123035

Fasina OO, Farkas BE, Fleming HP (2003) Thermal and dielectric properties of sweetpotato puree. Int J Food Prop 6(3):461–472. https://doi.org/10.1081/JFP-120021459

Flory S et al (2021) Increasing post-digestive solubility of curcumin is the most successful strategy to improve its oral bioavailability: a randomized cross-over trial in healthy adults and in vitro bioaccessibility experiments. Molec Nutr Food Res 65(24). https://doi.org/10.1002/mnfr.202100613

Gopalakrishnan K et al (2023) Applications of microwaves in medicine leveraging artificial intelligence: future perspectives. Electronics 12(5):1101. https://doi.org/10.3390/electronics12051101

Harwansh RK et al (2024) Recent advancements in gallic acid-based drug delivery: applications, clinical trials, and future directions. Pharmaceutics 16(9):1202. https://doi.org/10.3390/pharmaceutics16091202

Kazi M et al (2023) The development and optimization of lipid-based self-nanoemulsifying drug delivery systems for the intravenous delivery of propofol. Molecules 28(3):1492. https://doi.org/10.3390/molecules28031492

Li J et al (2024) Comparison of ultrasound and high-pressure homogenization emulsification: a promising fabrication strategy for total nutritional double emulsion-based product enriched with low-molecular-weight oyster peptides. LWT 212(August):116981. https://doi.org/10.1016/j.lwt.2024.116981

Liu J et al (2022) Fabrication of an aprepitant nanosuspension using hydroxypropyl chitosan to increase the bioavailability. Biochem Biophys Res Commun 631:72–77. https://doi.org/10.1016/j.bbrc.2022.09.031

Martínez RF, Cravotto G, Cintas P (2021) Organic sonochemistry: a chemist's timely perspective on mechanisms and reactivity. J Org Chem 86(20):13833–13856. https://doi.org/10.1021/acs.joc.1c00805

Meroni D, Bianchi CL (2022) Ultrasound waves at the service of photocatalysis: from sonochemical synthesis to ultrasound-assisted and piezo-enhanced photocatalysis. Curr Opin Green Sustain Chem 36:100639. https://doi.org/10.1016/j.cogsc.2022.100639

Qi K et al (2022) Sonochemical synthesis of photocatalysts and their applications. J Mater Sci Technol 123:243–256. https://doi.org/10.1016/j.jmst.2022.02.019

Shaikshavali P et al (2020) A simple sonochemical assisted synthesis of nanocomposite (ZnO/MWCNTs) for electrochemical sensing of Epinephrine in human serum and pharmaceutical formulation. Colloids Surf, A 584:124038. https://doi.org/10.1016/j.colsurfa.2019.124038

Tyukanko V et al (2023) Development of an ultrasonic method for the quality control of polyethylene tanks manufactured using rotational molding technology. Polymers 15(10):2368. https://doi.org/10.3390/polym15102368

Zhang K et al (2021) Hydrodynamic cavitation: a feasible approach to intensify the emulsion cross-linking process for chitosan nanoparticle synthesis. Ultrason Sonochem 74:105551. https://doi.org/10.1016/j.ultsonch.2021.105551

Zhang H, Yu S, Xu H (2022) Ultrasound-guided microwave ablation for symptomatic adenomyosis: more areas of concern for more uniform and promising outcomes. J Interven Med 5(3):122–126. https://doi.org/10.1016/j.jimed.2022.06.001

Zhang B, Stefanidis GD, Van Gerven T (2024) Can ultrasound replace seeding in flow reactive crystallization of an aromatic amine? Org Process Res Dev. https://doi.org/10.1021/acs.oprd.4c00385

The manufacturer's authorised representative in the EU is Springer Nature Customer Service Centre GmbH, Europaplatz 3, 69115 Heidelberg, Germany. If you have any concerns regarding our products, please contact ProductSafety@springernature.com

Printed and bound by CPI Group (UK) Ltd, Croydon, CR0 4YY
18/02/2026
02055611-0001